*A Ray of Light in a
Sea of Dark Matter*

PINP⬤INTS

Pinpoints is a series of concise books created to explore complex topics by explaining key theories, current scholarship, and important concepts in a brief, accessible style. Each Pinpoints book, in under 100 pages, enables readers to gain a working knowledge of essential topics quickly.

Written by leading Rutgers University faculty, the books showcase preeminent scholars from the humanities, social sciences, or sciences. Pinpoints books provide readers with access to world-class teaching and research faculty and offer a window to a broad range of subjects, for a wide circle of scholars, students, and nonspecialist general readers.

Rutgers University Press, through its groundbreaking Pinpoints series, brings affordable and quality educational opportunities to readers worldwide.

When complete, the series will comprise the following five volumes:

Deborah Carr, *Worried Sick: How Stress Hurts Us and How to Bounce Back*

Nicole Fleetwood, *On Racial Icons: Blackness and Public Imagination*

James W. Hughes and Joseph J. Seneca, *New Jersey's Postsuburban Economy*

Toby C. Jones, *Running Dry: Essays on Water and Environmental Crisis*

Charles Keeton, *A Ray of Light in Sea of Dark Matter*

A Ray of Light in a
Sea of Dark Matter

CHARLES KEETON

RUTGERS UNIVERSITY PRESS

New Brunswick, New Jersey, and London

Library of Congress Control Number: 2014930070

ISBN: 978-0-8135-6532-3 (web PDF)
ISBN: 978-0-8135-7212-3 (ePub)

Visit our website: http://rutgerspress.rutgers.edu

Manufactured in the United States of America

After they've clicked the light off
And everybody's said goodnight,
What's in the dark?
—Carl Memling

There are more things in heaven and earth, Horatio,
Than are dreamt of in your philosophy.
—William Shakespeare's *Hamlet*

CONTENTS

Preface

THE STORY OF COSMOLOGY is one of hubris and humility. Hubris to think that we humans can peer out from our perch on Earth and comprehend the totality of the universe. Humility to discover how insignificant we are in the cosmic scheme. When Copernicus displaced the Earth from the center of the universe, that was just the beginning. Now we understand that our star is one of hundreds of billions in our galaxy, which itself is one of hundreds of billions of galaxies in the universe. Worse, the substance of which we are made—and the subject of all scientific inquiry until the twentieth century—turns out to be as exceptional as dewdrops on a spider's web. The underlying fabric of the universe is something altogether different, unknown and exotic.

And yet ... the humility is surely suffused with awe for what we have learned. In our galaxy, the stars and gas that we see are embedded in an enormous halo of matter that we do not see. The "dark" matter surrounds us and penetrates us; pieces of it are flying through you at this moment. It also binds our galaxy together; the Milky Way, like most spiral galaxies, is spinning so fast that it would fling itself apart if not for the gravitational pull of dark matter. Throughout the universe, dark matter outweighs the familiar visible matter by a ratio of 5:1, and it creates the cosmic web on which visible galaxies hang. Truly there is far more in the heavens than we ever dreamed.

Our knowledge of dark matter seems quite remarkable given that we have never seen the stuff. It derives from that special blend of curiosity, commitment, creativity, and critical thinking known as scientific inquiry. To many scientists, research has an air of childlike wonder; it begins with such eternal questions as, What's this? Why did that happen? What's over here? What would happen if . . . ? Playful curiosity is not enough, however. Scientists need to be creative in developing hypotheses and designing experiments to test them. They need to be rigorous in refining ideas that pass the test and discarding those that don't. Last, but certainly not least, they need to be able to sort through messy, incomplete, and even contradictory evidence to extract appropriate conclusions.

The study of dark matter, in other words, requires intellectual skills that transcend cosmology and speak to the broad capabilities of the human mind. Our ability to marshal such skills to discover what's in the dark suggests that our cosmic hubris is, perhaps, not entirely mistaken. We can in fact peer into the depths and understand what the universe is made of, when we put our minds to it.

This book recounts some of the intellectual history of dark matter. We focus on how astronomers use light rays as a diagnostic tool to study material that would otherwise remain hidden. The story begins a century ago, when Albert Einstein predicted that gravity bends light. He did not anticipate the scope of that simple idea; a half century passed before the Norwegian astrophysicist Sjur Refsdal recognized the cosmological potential of light bending, and another three decades before the Hubble Space Telescope and other instruments turned gravitational lensing into a tool for cosmology. Today, lensing plays a key role in the quest to understand dark matter.

This book draws on my research in gravitational lensing, and my teaching in astrophysics more broadly. My colleagues at Rutgers University and around the world have contributed a great deal to my ongoing education. They are too many to name individually, but some of their work is cited in the "Notes on Sources." This is not a textbook but it does contain some specialized terminology that may not be familiar, so terms in **boldface** are defined in the glossary. I am grateful to Charles Bergquist, Art Congdon, Charley Keeton, Allan Moser, Leonidas Moustakas, Kelly Wieand, and my editor Leslie Mitchner for valuable comments on the manuscript. The custom lens shown in chapter 2 was built by William Schneider and the machine shop in the Department of Physics and Astronomy, with help from Jack Hughes, Dave Maiullo, and Kathleen Sindoni. My work has been supported by two grants from the National Science Foundation (AST-0747311 and AST-1211385).

*A Ray of Light in a
Sea of Dark Matter*

CHAPTER I

What's in the Dark?

ASTRONOMERS HAVE LONG BEEN denizens of the dark, spending nighttime hours on remote mountaintops capturing tiny traces of light from afar. For decades, observers had to ride on telescopes all night, cramped in a cage opposite the mirror, to keep the instrument focused on the target during long-exposure photographs. Then they would carry the photographic plates to the darkroom to be developed. Later, teams of assistants would pore over the plates to measure planets, stars, and galaxies. Observational astronomy was painstaking, and every **photon** was precious.

A practical change is under way, driven (as in so many other fields) by technology. Today, astronomers guide telescopes by computer from control rooms that, while perhaps not plush, at least offer light, heat, and Internet access. They record data in electronic files that are easily copied and shared. With some telescopes, observers can do everything remotely, working from their offices or even living rooms. (There is ongoing debate about the relative costs and benefits of being on site for observing.) One thing remains the same: photons are still precious, even if they are collected in new ways.

A philosophical shift is also afoot: from working *in* the dark to working *on* the dark. Beginning in the 1970s, astronomers realized there is more matter in the universe than meets the eye. Observers started looking for ways to map the dark matter, theorists studied how it would affect the growth of structure in the universe, and particle physicists got excited about the prospect of finding entirely new kinds of particles. Later, in the 1990s, observers found strong evidence for an even more exotic substance dubbed "dark energy." A plethora of measurements now indicates that dark matter and dark energy together compose 95 percent of the stuff in the universe. While science has always addressed the unknown, cosmology now has to deal with the unseen.

The challenge is acute because astronomy is an observational science. We cannot do experiments on galaxies, let alone the universe; all we can do is look around, search for patterns, and invent ideas that might explain what we see. A good lesson comes from early work on stars. In the late nineteenth and early twentieth centuries, Harvard College Observatory hired a number of women to help process astronomical data. (The women came to be known as "computers.") Williamina Fleming analyzed measurements known as **spectra** in which starlight is spread out like a rainbow to show all of the different colors. Dark bands appear where certain wavelengths of light are absorbed by atoms or molecules, and Fleming introduced a classification scheme based on the patterns she saw. Later, Annie Jump Cannon discovered a connection between the absorption pattern and temperature of a star. That relation made it possible for Ejnar Hertzsprung and Henry Norris Russell to plot the luminosities of stars versus their temperatures and learn that most stars fall on a sequence running from bright hot blue stars down to dim cool red stars. The strong pattern in the **Hertzsprung-Russell diagram** inspired research on the physics of stars and ultimately led to the discovery that the sequence in temperature and brightness is really a sequence in mass: more massive stars have stronger gravity that squeezes the gas to create higher temperatures and luminosities. This example shows that with a critical eye to spot patterns, and a creative mind to interpret them, we can learn a lot about objects we cannot touch.

What about material we cannot even see? Here we turn to Isaac Newton, who discovered a deep connection between motion and mass. When he thought about general forms of motion, Newton deduced three fundamental laws:

1. Left alone, an object will stay at rest or move in a straight line at a constant speed.
2. An object can change speed or direction only if it is acted on by an external force; the change depends on the strength and direction of the applied force.
3. If object A exerts a force on object B, then object B exerts an equal and opposite force back on object A.

According to these principles, an apple falls from a tree not because it "wants" to be on the ground but because some unseen force impels it. The Moon orbits the Earth not because it "wants" to move along a circle or ellipse but because an unseen force pulls it away from straight-line motion. That idea was already profound; no one had previously thought about an invisible force transmitted through empty space.

But Newton went further. Using his second law of motion, he computed the amount of force that must act on the apple or the Moon to generate the motion. Newton then asked, What if it is the *same* force affecting both objects? The numbers worked out if he assumed that gravity scales inversely with the square of the distance between two objects. (If the objects move twice as far apart, gravity weakens by a factor of $2 \times 2 = 4$.) Even better, applying the inverse square force law to planets orbiting the Sun led to predictions that matched the laws of planetary motion that Johannes Kepler had extracted from observational data. Talk about "Eureka!": in one fell swoop, Newton explained centuries' worth of cosmic motion measurements.

These principles can be extended to detect mass that was previously unknown. In 1781, William Herschel discovered the object now known as Uranus. Its motion was initially found to be consistent with Newton's extrapolation of Kepler's laws, so it was identified as a planet orbiting the Sun. As the measurements improved, the motion was found to deviate slightly from Newtonian predictions. Urbain Le Verrier and John Couch Adams separately analyzed the deviations and argued that they could be caused by gravity from another planet beyond the orbit of Uranus. Sure enough, in 1846 Johann Galle and Heinrich d'Arrest used Le Verrier's predictions to discover the planet Neptune.

Similar logic now pervades astrophysics. When we see something move, we can use Newton's laws of motion to find the force that induces the motion. We can then use Newton's law of gravity to figure out how much mass is needed to create the required force. This connection between motion and mass is one of the pillars of modern astrophysics. As long ago as the 1930s, Fritz Zwicky noticed that galaxies in collections known as **clusters of galaxies** move much faster than they should if gravity comes only from the galaxies themselves. His analysis was ahead of its time, however. The idea that fast motions reveal "missing mass" began to gain traction in the 1970s and 1980s when Vera Rubin and others realized that the speeds of stars in spiral galaxies require much more gravity than can be attributed to the visible matter (see chapter 3). Today, astronomers use the **motion/mass principle** to discover planets orbiting other stars, find monstrous black holes at the centers of galaxies, and much more.

There is a catch. First, we cannot always measure motion in full detail. For example, we can determine how fast stars in a galaxy are moving toward or away from us, but we cannot detect how fast they are moving left/right or up/down in the plane of the sky. Given the uncertainties, we might be able to find multiple configurations of mass that could give rise to the observed motion. It is important to recognize the uncertainties, see if we can find ways to resolve them, and deal with them if we can't.

Second, we have to guard against being misled. The connection between motion and mass runs through Newton's laws of motion and gravity. If the laws themselves are incorrect, the conclusions will be erroneous (even if the logic is sound). Urbain Le Verrier's success with Uranus was actually complemented by failure of this sort with Mercury. Kepler found that planet orbits are ellipses, and Mercury's is the most elongated of the eight major planets. If Mercury were the only planet orbiting the Sun, it would trace the same ellipse over and over forever. In fact, Mercury's ellipse precesses, or shifts slightly from one orbit to the next. Most of the precession can be attributed to gravitational influences from the other planets—but not all. Le Verrier argued that the additional, unexplained motion is caused by an unseen planet closer to the Sun, which he named Vulcan. Inspired by this prediction, a number of amateur astronomers claimed to see a new planet crossing the face of the Sun, and Le Verrier died believing his prediction had been confirmed. The observational claims could not be verified, however.

Mercury's excess precession remained unexplained until Albert Einstein published his general theory of relativity in 1915. According to general relativity, the effects of gravity do not quite match Newton's inverse square law; there are small deviations close to a massive object like a star. Einstein discovered that those deviations are exactly what is needed to explain Mercury's orbit. In other words, Mercury is not like Uranus: one case revealed new mass in the solar system, while the other actually indicated changes in the laws of physics.[1]

Some skeptics have argued that dark matter is the new Vulcan—a misguided conclusion drawn from a misapplication of Newton's (and Einstein's) laws. They suggest that puzzling motions in the cosmos might indicate not new matter but new physics. In 1983, Mordehai Milgrom introduced **Modified**

Newtonian Dynamics (MOND), in which Newton's second law of motion is adjusted in a way that could reproduce the motions of stars in galaxies (without affecting motions on Earth or in the solar system; see chapter 3). Proponents argue that MOND is simpler than dark matter when it comes to explaining galactic rotation. They also point out that Newton's laws were thought to be immutable until relativity came along; are we certain that relativity itself can never be modified?

Despite those arguments, most astronomers are firmly convinced that dark matter is real. The reasoning has a number of facets. First, we understand how Newton's laws break down in the regimes where Einstein's laws kick in, so we know when it is safe to stick with classical mechanics and when we need to shift to relativity. Second, Einstein's theory has passed every test to which it has been subjected. Reasoning based on Newton's and Einstein's laws therefore seems as if it should be sound. Third, some specific astronomical systems are difficult to understand without dark matter (see the discussion of the Bullet Cluster in chapter 6).

A fourth facet involves a conceptual principle known as **Ockham's razor**. William of Ockham was a medieval theologian and philosopher who sought to make scholarly reasoning more efficient. Today many people interpret his thinking in terms of simplicity, formulating a principle that might be stated as, "The simplest explanation is probably the correct one." A more nuanced interpretation allows that ideas may be intrinsically complex but still advocates parsimony: "When faced with competing hypotheses that are otherwise equal, pick the one with the fewest new assumptions." Ockham's razor, in other words, should be used to trim away ideas that are needlessly complex. This is not a rigorous scientific principle; experimental evidence remains the arbiter of truth in science. But it is a rule of thumb that can be helpful when developing new hypotheses or models and performing an initial assessment.

Dark matter survives Ockham's razor because it represents a single concept that explains a myriad of astronomical observations. It explains the motions of stars in galaxies, the properties of galaxy clusters, the distribution of galaxies in the universe, the overall expansion of the universe, and much more. Without dark matter, in fact, it is hard to make sense of the vast array of cosmological data now available. There remain some puzzles, to be sure, but they are considered to be fodder for ongoing research rather than reasons to discard the idea of dark matter altogether. As a competing hypothesis, MOND does well with galaxies but has some trouble with other systems. There are enough ambiguities that we cannot say for certain whether MOND fails to explain clusters of galaxies and the universe as a whole, but even proponents admit that MOND alone is not enough; some mass beyond the visible stuff is needed (although it might be something we know about, such as **neutrinos**).

A fifth facet of the belief in dark matter comes from particle physics. While the **Standard Model of particle physics** describes all the particles we know now, theorists have long suspected that other particles could exist. In the 1960s, for example, work by Peter Higgs led to predictions of a new particle that was ultimately discovered at the Large Hadron Collider in 2012–13. (The predictions were honored with the 2013 Nobel Prize in Physics.) That is not the end of the story, however, by a long shot. Many other types of particles have been hypothesized for a variety of reasons, and some of them could have the right properties to be cosmic dark matter (see chapter 4).

Such arguments have convinced the vast majority of astronomers and physicists that what's in the dark is a new form of matter. Assuming that to be true, what do we know about the substance? A little or a lot, depending on your perspective. Although we do not actually know what it is, we have begun to figure out what it is like; we know, in other words, few nouns but many adjectives. First, dark matter is, well, *dark*; it does not emit, absorb, or reflect light. That suggests it is electrically *neutral*, because charged particles interact with light through the electromagnetic force. Dark matter is *abundant*; in the universe as a whole, it outweighs normal matter by a factor of five. Dark matter is *pervasive*; it is found in many different types of galaxies, in clusters of galaxies, and throughout the universe. But dark matter is not distributed uniformly; rather, it is *clustered* so the density is higher in some places than in others. This is caused by gravity; any region of the universe that has a little more than the average amount of matter will have a little extra gravity, so it will tend to draw in even more matter from surrounding regions.

The fact that dark matter is clustered suggests that it is "*cold*," a term physicists use to indicate that the material is slow compared with the speed of light. If the particles were fast, they would fly away instead of collecting in **halos** around galaxies and clusters of galaxies. Finally, dark matter appears to be *exotic*—some fundamental particle that is different from all of the particles we have seen in physics experiments to date.

We might view all of these adjectives as pieces of a jigsaw puzzle that, when complete, will show what dark matter is. Unfortunately, we do not know how the pieces fit together, or how many remain to be found. So we do our best to assemble a plausible picture and then look for gaps we can try to fill in. The conventional picture today suggests that dark matter particles weigh a few tens or hundreds of times as much as a proton. If so, the particles are all around you; there are a few dozen in every liter of volume throughout the solar system and even on Earth. The particles are zipping through the room in which you are sitting at several hundred thousand miles per hour. They interact so weakly that they rarely hit anything, but every once in a while a piece of dark matter might collide with a piece of normal matter on Earth. Physicists are now trying to use such collisions to catch the particles and measure their properties (see chapter 4). Particles of dark matter might also be able to interact with one another, annihilating into a cascade of normal particles.

These two types of collisions provide opportunities to probe dark matter in new ways. They have some limitations, however. It goes without saying that collisions with particles on Earth probe dark matter right here. We can assume that dark matter here is the same as dark matter everywhere, but experiments on Earth cannot test this assumption. Collisions among dark matter particles might occur throughout the universe, but they would be most likely in places where the density of dark matter is particularly high. Also, particles created by dark matter collisions might be hard to distinguish from particles produced by more mundane astrophysical processes. Several recent studies have in fact found excess numbers of the types of particles that could be created by dark matter collisions, but it is unclear whether those particles come from dark matter or known sources. Finally, the whole notion that dark matter particles can collide with each other or with normal particles depends on certain assumptions about the particles. It is entirely possible that dark matter will play hard to get (even more than it has so far) by having properties that render it undetectable in the lab.

Astrophysics therefore remains a central part of the quest to understand dark matter. Even so, it would be nice to develop new ways to investigate the substance. Here we turn to Einstein, who discovered that the motion/mass principle applies not only to material objects like planets and stars but also to light rays. As we will see in chapter 2, general relativity predicts that gravity bends light rays such that a massive object can act as a special kind of lens—a "gravitational" lens—for light from a background source. Analyzing the motion of light that passes near a galaxy is akin to analyzing the motion of stars within the galaxy; both let us measure mass even if it is invisible. As an extension of the motion/mass principle, light bending has become a versatile tool for astrophysics and cosmology. In the rest of this book, we will see examples of the creative ways that astrophysicists harness the power of gravitational lensing to study dark matter.

The work we will encounter serves as a microcosm of astrophysics as a whole, and even a broad swath of scientific inquiry, in terms of the intellectual skills that are integral to success. Astrophysicists need technical skills to carry out observations and calculations that provide the fundamental facts. But they also need the critical thinking skills of judgment, logic, reading, and writing in order to evaluate evidence, construct arguments, communicate their own ideas, evaluate their colleagues' ideas, and much more. They also need creativity in order to see connections (for example, between a falling apple and planets) and to develop hypotheses that may explain the facts (to guess at the picture on the jigsaw puzzle, in other words). And, critically, they must be willing to explore ideas, follow them to their logical conclusions, and discard those that prove to be incorrect. Beyond discussing dark matter itself, this book aspires to illustrate the inquiry that has led to the understanding we have now.

Let me note that this book does not address what has turned out to be the largest component of the universe: dark energy, a bizarre substance that is causing the expansion of the universe to accelerate. Dark matter is plenty to keep us busy here; dark energy is discussed in other books. For example, *The Extravagant Universe: Exploding Stars, Dark Energy, and the Accelerating Cosmos* by Robert Kirshner is written by a member of one of the research teams recognized by the 2011 Nobel Prize in Physics; and *Einstein's Telescope: The Hunt for Dark Matter and Dark Energy in the Universe* by Evalyn Gates describes the role that gravitational lensing has played in research on dark energy.

CHAPTER 2

When Mass Is Like Glass

BEFORE DISCUSSING HOW ASTRONOMERS use light bending to investigate mass, it is good to understand the physical phenomenon itself. The term "gravitational lensing" is very apt, because mass actually acts like a certain kind of glass lens. Starting with traditional optics can help clarify what happens with gravity.

When a light beam passes from air into glass, it slows down. (The same can be said of water, oil, or any other translucent substance that is denser than air.) If the beam hits the surface at an angle, one side slows down before the other, causing the beam to bend (see figure 2.1). This phenomenon is known as "refraction," and it is prevalent throughout optics. Isaac Newton used it to discover that white light contains all of the colors of the rainbow. The angle of refraction depends on the wavelength (or color) of light, so a beam of white light passing through a prism gets spread into its component colors. Rainbows in the sky occur when sunlight refracts through water droplets in the air.

Our eyes use refraction to help us see. The lenses in our eyes have a curved surface, so different light rays bend by different amounts depending on where they hit the lens. If the surface is convex (as shown in figure 2.2), the light rays all bend toward each other, and the lens is called a "converging lens." If the curvature of the surface is just right, all the rays from a distant source come together ("focus") at the same point. Ideally, the focal point lies on the retina, forming an image the brain can

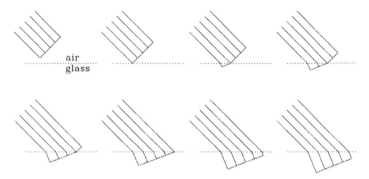

2.1. A light beam approaches a pane of glass at an angle. As it enters the glass, the light slows down, causing the beam to refract.

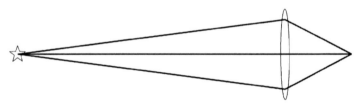

2.2 A lens with a convex surface causes light rays to converge. Here the rays come to a focus on the right-hand side of the lens.

process. Muscles in the eye can change the shape of the lens to focus on objects at different distances.

For people (like me) who are nearsighted, the lens in the eye cannot focus properly. It needs a little help from an artificial lens with a different shape. If the surface is concave (as shown in figure 2.3), the light rays all spread apart, and the lens is called a "diverging lens." How does this help me see? If the curvature is just right, all the rays appear to have originated from a point that is much closer than the original source. My eyeglasses, in other words, manipulate light rays so that it looks like a source is close enough for my nearsighted eyes to bring into focus.

In figures 2.2 and 2.3, notice that the light bends toward the thicker part of the lens. What happens if we mix things up and

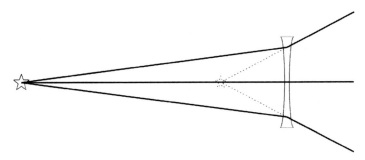

2.3 A lens with a concave surface causes light rays to diverge. Here the rays seem to come from a point closer to the lens (indicated by the dotted star).

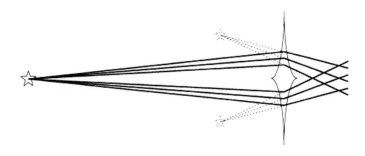

2.4 If the surface of the lens is concave and thick in the middle, light rays diverge and bend toward the center. Even though there is just one source on the left, the rays on the right appear to have originated from different points (indicated by dotted stars).

make the surface concave but thick in the middle? (See figure 2.4.) First consider light rays that pass through the top half of the lens. The rays diverge because the surface is concave, but they bend downward (toward the thick region). As a result, these rays all appear to emanate from a point above the middle of the lens. Now consider the rays that pass through the bottom half of the lens. Again they diverge, but these bend upward and appear to originate at a point below the middle. As a result, an observer on the right-hand side sees light rays that appear to come from two

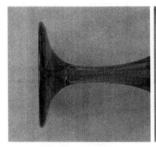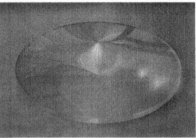

2.5 The left panel shows that the base of a wine glass has a shape similar to the lens depicted in Figure 2.4. The right panel shows a special glass lens built with a similar shape by William Schneider and the machine shop in the Department of Physics and Astronomy, with help from Jack Hughes, Dave Maiullo, and Kathleen Sindoni.

different points. The lens makes it look like there are two sources of light, when in fact there is just one.

While this may seem abstract, a familiar object actually looks a lot like the lens we are considering: the base of a wine glass (see figure 2.5). We will see below that such a lens can bend light to make a candle flame look like two or more flames, or even a complete ring. These curious phenomena occur because (a) light rays bend toward the center of the lens, and (b) light rays that are closer to the center bend more, while light rays that are farther from the center bend less.

Gravity creates the same two effects. When Albert Einstein developed his general theory of relativity, he pictured gravity in terms of the curvature of spacetime. (In relativity, space and time are connected and cannot be discussed separately; spacetime is the joint structure.) Mass causes spacetime to curve, similar to the way a heavy ball causes a rubber sheet to bend and stretch (see figure 2.6). Any object (including a light ray) that moves through this curved spacetime must follow a curved trajectory. The effect is much smaller than depicted in the figure; Einstein calculated that a light ray grazing the surface of the Sun would

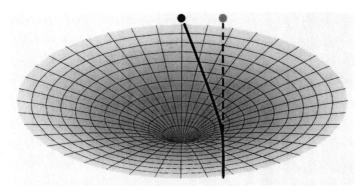

2.6 In Einstein's theory of relativity, spacetime is distorted in the vicinity of a massive object. The black point shows a distant light source, and the solid line shows the path of a light ray from it. The gray point shows that the light appears to have originated from a different position.

be bent by an angle that is a thousand times smaller than the diameter of the Moon as seen from Earth. Normally such light rays are swamped by light from the Sun itself, but they become visible when the Moon blocks the Sun's light during a solar eclipse. Arthur Eddington and Frank Dyson led expeditions to observe an eclipse in 1919 and found that the apparent positions of background stars indeed shifted by an amount consistent with Einstein's prediction. The *New York Times* immortalized the discovery with the headline "Lights All Askew in the Heavens" (November 10, 1919).

If the Sun were much more compact (so it would not block light from background stars), it could act just like the wine glass. Figure 2.7 shows that the analogy works because gravity is an attractive force (so light rays bend toward the mass M), and the force weakens with distance (so light rays farther from M bend less). Mass, in other words, can act just like glass to create two or more images of a background source.

Today we can use Einstein's theory to predict what gravitational lensing looks like. Figure 2.8 demonstrates that both glass

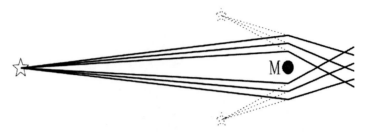

2.7 Light passing a massive object can bend in much the same way as light going through the glass lens in figure 2.4.

and gravitational lenses produce three distinct effects. Each lens creates two images of the source, which lie on opposite sides of the center. The images are shifted relative to the source because the light rays are deflected. Also, the two images have different shapes and brightnesses. A fourth effect is shown in figure 2.9. A source directly behind the lens creates a full circle known as an **Einstein ring**. In essence, there are images on both sides of M, but when the observer, lens, and source are perfectly aligned, the whole problem can be rotated so "both sides" becomes "all sides."

Einstein figured all this out, but not until two decades after he originally predicted light bending. In fact, he did so only at the behest of a Czech engineer named Rudi Mandl, who argued that a foreground star might be able act as a lens for a background star. Einstein dutifully published the calculations depicted in figures 2.8 and 2.9, but he was not enamored of the idea. He concluded the paper by remarking, "There is no great chance of observing this phenomenon." To absolve himself further, Einstein wrote in a cover letter to the journal editor, "Let me also thank you for your cooperation with the little publication, which Mr. Mandl squeezed out of me. It is of little value, but it makes the poor guy happy."

In this judgment, Einstein was both right and wrong. He was correct that the chance of seeing one star lens another star is very small: less than one in a million, in fact. What Einstein

2.8 The left panel shows a candle flame when viewed through the glass lens shown in figure 2.5. The body of the candle is visible at the bottom of the image toward the left, and at the top of the image toward the right (because this image is inverted). The right panel shows a theoretical prediction of a similar image configuration produced by a gravitational lens.

2.9 Illustration of Einstein rings produced by glass (left) and gravitational (right) lenses.

did not anticipate was that, some six decades later, astronomers would develop the motivation, commitment, and technology to observe millions of stars over and over again in order to see rare lensing events. As discussed in chapter 4, Einstein's "little publication" came to play a big role in the discovery that dark matter is not just matter that is just difficult to see; it must be something unfamiliar.

Einstein's focus on stars revealed something of a lack of imagination when it came to gravitational lensing. Fritz Zwicky was much more comfortable thinking outside the box. He left a trail of ideas that turned out to be decades ahead of their time, including both gravitational lensing and dark matter. The year after Einstein's paper appeared, Zwicky pointed out that "extragalactic nebulae [what we now call galaxies] offer a much better chance than stars for the observation of gravitational lens effects." His idea lay fallow for almost three decades because people did not know what kinds of objects might serve as light sources to be lensed by galaxies. The situation changed in the 1960s with the discovery of quasi-stellar radio sources, or **quasars**. These objects look like stars (hence "quasi-stellar") but lie so far away that they must be shining with the brightness of an entire galaxy. They were originally identified in the radio portion of the electromagnetic spectrum, but many emit significant amounts of visible light as well. Around the same time, Sjur Refsdal pointed out that gravitational lensing of distant sources such as quasars and exploding stars called supernovae could be very valuable for cosmology.

Even with such good sources, the probability of having one aligned with a massive galaxy is low. It was 1979 before Dennis Walsh, Robert Carswell, and Ray Weymann found the first gravitational lens system (known formally as Q0957+561, based on its coordinates in the sky). They saw two quasars whose spectra were so similar that the objects had to be either twins or lensed images of a single source. Peter Young and his collaborators confirmed the lensing interpretation when they found the galaxy responsible for bending the light from a background quasar.

Additional discoveries followed quickly as more quasars were observed. The second system found, PG 1115+080, was puzzling because it seemed to show three images rather than the expected two. Follow-up observations resolved one of the images into two, meaning that gravitational lensing produces a total of four images of the quasar.

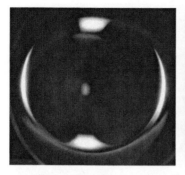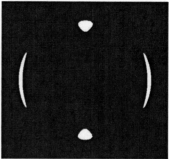

2.10 Illustration of four-image configurations produced by glass (left) and gravitational (right) lenses.

We now understand that four-image lensing is quite normal. Few galaxies have the perfect spherical symmetry needed to produce complete Einstein rings. It is more common for galaxies to have an elliptical shape with two lines of symmetry: an ellipse looks the same if reflected about its long or its short axis. A source directly behind such a lens creates two images on each axis. Other positions can create other configurations with four images, as shown in figure 2.10. Today, several dozen quadruply-imaged quasars have been found and observed with the Hubble Space Telescope (see figure 2.11).

One additional phenomenon can be seen in lensed quasars. The images correspond to rays that take different paths through the universe and thus have slightly different lengths. Also, the rays go through different parts of the lens's gravitational field, which means that they experience different amounts of a relativistic effect known as "gravitational time dilation." As a result, light takes a little longer to travel some of the paths than others. If the source gets brighter or fainter, the changes will appear in different images at different times. The **time delays** between the images are typically a few weeks or months, which is convenient if we want to measure them, but remarkably small given that the overall light travel time can approach ten billion years.

2.11 Hubble Space Telescope image of the four-image gravitational lens system SDSS J0924+0219. Credit: Keeton et al. (2006). © AAS. Reproduced with permission.

If the light source is extended rather than compact, it can be stretched into a partial or complete Einstein ring (see figure 2.12). The lensing can be even more dramatic if the lens is not a galaxy but an entire cluster of galaxies. These are the largest gravitationally bound structures in the universe, containing hundreds or even thousands of galaxies all moving in an enormous halo of dark matter that may weigh as much as a quadrillion Suns. Galaxy clusters can produce some of the most visibly striking manifestations of gravitational lensing, as shown in figure 2.13. These systems can also serve as "cosmic telescopes" that boost

2.12 Hubble Space Telescope images of Einstein ring gravitational lens systems. Credit: NASA, ESA, A. Bolton, and the SLACS team.

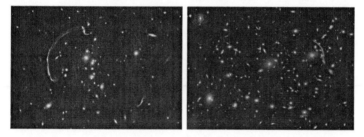

2.13 Hubble Space Telescope images of gravitational lens systems created by clusters of galaxies. Most of the bright blobs are galaxies orbiting each other in a common halo of dark matter. The long arcs are gravitationally lensed images of background galaxies. Credit (left): NASA, ESA, J. Rigby, K. Sharon, M. Gladders, and E. Wuyts. Credit (right): NASA, ESA, the Hubble SM4 ERO Team, and ST-ECF.

our ability to peer into the depths of cosmic time and study some of the first galaxies to form in the universe (see chapter 6).

From multiply-imaged quasars to giant arcs and rings, observed lens systems appear to be quite diverse. Yet they all arise from three gravitational effects: deflection, distortion, and delay. The three D's of lensing give astronomers a variety of ways to learn about matter in the universe.

CHAPTER 3

How Do You Weigh a Galaxy?

As MENTIONED IN CHAPTER I, astronomy often begins with the identification of patterns. The patterns we see first are not always the ones that have the most physical significance, however. It is important to understand which patterns are coincidental and which offer deep insights into the structure of galaxies.

Stars are scattered across the night sky, but our minds try to impose some order on the randomness. Human societies have identified two types of patterns in the stars. Asterisms are groupings of a few stars that seem to take on recognizable shapes: the Big Dipper, the Summer Triangle, and the Teapot, to name but three. Constellations are more complex structures that often have stories attached to them. The International Astronomical Union recognizes eighty-eight constellations, and the ones in the northern sky are mostly associated with figures from Greco-Roman mythology: Orion the Hunter, Taurus the Bull, Cassiopeia the Queen, and so on. Travelers have long used constellations to navigate, farmers have used them to mark the seasons, and astronomers have used them to map the sky. Nevertheless, we now understand that these apparent patterns do not have any physical significance. The stars in a given constellation are not actually associated with each other; they often lie at different distances and just happen to appear close to each other as viewed from Earth. Observers in different parts of our galaxy would see different patterns. For that matter, the patterns we see will change over millennia as the stars move through space.

Running across the sky is a band of diffuse light known as the Milky Way. It is difficult to see now because of light pollution, but it is very prominent when the sky is dark. While the Milky Way appears to have a smooth distribution, Galileo used an early telescope to discover that it is actually a collection of innumerable faint stars. Sprinkled among the stars are fuzzy objects called **nebulae** (the singular is *nebula*, from the Latin for "cloud"). Some of these have irregular shapes. Others are round and seem to resemble planets, so they became known as "planetary nebulae." Still others have a clear spiral structure, and hence are known as "spiral nebulae." In the 1770s, Charles Messier assembled a catalogue of diffuse objects, including many nebulae, that is still in use today. (Messier actually viewed these objects as annoyances—distractions from comets, which were his main interest. He created the catalogue as a way to keep track of things he had already seen and decided were not comets.)

By the early twentieth century, some people began to suggest that spiral nebulae are separate "island universes." They invoked structural patterns, arguing that the milky band in the sky is what we would see if Earth were located inside its own spiral nebula, and conversely that spiral nebulae show what the Milky Way would look like if seen from the outside. Others dismissed the notion, saying that if spiral nebulae were outside the Milky Way, they would have to be unthinkably large in order to be visible. More specifically, Adriaan van Maanen claimed to see one spiral nebula rotate and argued that the speed would be impossibly fast if the nebula were as big as a galaxy. The competing viewpoints were presented in a famous 1920 debate between Heber Curtis (who supported the "island universe" hypothesis) and Harlow Shapley (who opposed it). The "Great Debate" carries historical weight because it shows prominent scientists grappling with evidence that was incomplete and even incorrect.[1] Yet it did not settle the issue. That happened only a few years later, when Edwin Hubble used a special type of variable star to measure the distances to several spiral nebulae, finding that they are indeed far enough to be separate galaxies.[2]

Continued observations demonstrated that stars in a **spiral galaxy** occupy a thin, rotating disk. The thinness can be seen directly in galaxies that happen to be oriented so that they appear edge-on. If the rotation cannot actually be seen, you may wonder how it can be discerned. Think of a train whistle. The whistle seems to have a higher pitch when the train is coming toward you than when it is moving away. When the train approaches you, the sound waves get bunched together by the train's motion and therefore seem to have a higher pitch; when the train recedes, the sound waves get spread apart and seem to have a lower pitch. This is called the **Doppler effect** for sound. A similar effect happens for light, although instead of pitch we talk about color. Light appears to have a bluer color when the source is moving toward you, and a redder color when the source is moving away. Doppler measurements of galaxies reveal that stars on one side of the disk are moving toward us while stars on the other side are moving away, and the typical speeds are a few hundred kilometers per second (hundreds of thousands of miles per hour).

These measurements provided the first opportunity to weigh galaxies using the motion/mass principle. As discussed in chapter 1, measuring motion lets us determine how much mass is needed to generate it. Even better, finding patterns of motion lets us study how the mass is distributed. In our solar system, for example, almost all the mass is concentrated in the center, so planets farther from the Sun feel less gravity and hence move more slowly. (Think of a ball on a string: swinging the ball hard makes it go fast, but swinging it gently makes it move more slowly.) In a galaxy, the pattern of motion may be more complicated because the mass is spread throughout the disk. To predict the **rotation curve**, which shows the speed of stars as a function of distance from the center, we need to convert the light we see into the mass that generates motion. The conversion depends on the population of stars in a galaxy, which we may not know in detail. If we assume the population is relatively uniform throughout a galaxy, we predict a rotation curve with the form shown by the dotted line in figure 3.1. The key feature is that stars near the edge of a galaxy are expected to move slower than stars closer to the center because they are farther from the bulk of the mass.

When Vera Rubin, Morton Roberts, and others began measuring rotation curves in the 1960s and 1970s, they found a surprise: the stars on the edge move just as fast as the stars closer in. Observed rotation curves, in other words, are nearly flat (as shown by the solid line in figure 3.1). The implication is that stars twice as far from the center of a galaxy feel about twice as much mass as the closer stars, even though there is not twice as much light. Apparently galaxies are filled with mass we cannot see.

How can we account for the "missing mass"? One possibility is that it is hidden in the disk. If the stellar population is not actually uniform, the outer parts of the galaxy could be filled with dim stars that contribute a lot of mass but little light (such as brown dwarfs; see chapter 4). Yet that is not the only potential explanation. In the mid-1970s, some people began to suggest that spiral galaxies could be surrounded by extended, spherical **halos** of matter that is not seen. Rubin and her collaborators considered both hypotheses when interpreting observed rotation curves in 1978. About the halo hypothesis, they wrote, "Such models imply that the galaxy mass increases significantly with increasing r [distance from the center] which in turn requires that rotational velocities remain high for large r. The observations presented here are thus a necessary but not

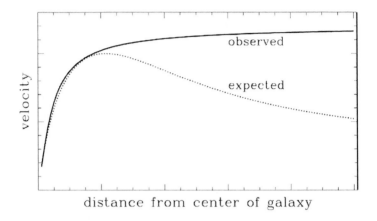

distance from center of galaxy

3.1 Schematic diagram of a spiral galaxy rotation curve. The plot shows the speed at which stars move in the galaxy, as a function of distance from the center of the galaxy. The dotted curve shows what would be expected if galaxies contain only the visible matter, while the solid curve shows the typical shape of observed rotation curves.

sufficient condition for massive halos." Notice the careful (and self-effacing) logic here: finding small rotation speeds could have disproved the halo hypothesis, but it does not necessarily follow that finding large speeds proves the hypothesis. In fact, the measurements could be explained equally well with extra matter in a spherical halo or in the disk itself. Rubin and colleagues were very frank in admitting that "the choice between spherical and disk models is not constrained by these observations." In other words, stellar motions imply that spiral galaxies contain a lot of unseen mass but do not tell us precisely where it is located.

That, at least, is the conventional conclusion, which follows from interpreting motion in the context of Newton's laws of motion and gravity. In 1983, Mordehai Milgrom introduced a third option. He suggested that instead of adding mass to make Newton's law of gravity produce more force, we could modify Newton's second law of motion so a given force generates more acceleration. Specifically, Milgrom proposed to keep Newton's second law as $F = ma$ when the acceleration due to gravity is as large as it is in the solar system ($a = 9.8$ m/s^2 on Earth, and $a = 0.006$ m/s^2 for Earth orbiting the Sun), but to make F proportional to a^2 when the acceleration is much lower ($a = 2 \times 10^{-10}$ m/s^2 for the Sun orbiting the Milky Way). He showed that this change could produce flat rotation curves and explain an observed relation between galaxy luminosities and rotation speeds, all without needing unseen matter.

All three hypotheses—disk dark matter, halo dark matter, and Milgrom's Modified Newtonian Dynamics—offer viable interpretations of spiral galaxy motions. To distinguish between them, we need new ways to weigh galaxies. Many have been developed over the years, but two are worth highlighting. One approach is to find objects other than individual stars that feel a galaxy's gravity. **Globular clusters** and **dwarf galaxies** are collections of stars (between a few hundred thousand and a few hundred million) that form coherent structures orbiting galaxies. Their motions toward or away from us can be measured with the Doppler effect, but their side-to-side motions are more difficult to determine. Even so, applying the motion/mass principle to globular clusters and dwarf galaxies suggests that mass continues to grow even beyond the visible edge of a galaxy.

Another approach is to examine other types of galaxies. **Elliptical galaxies** have smooth distributions of stars (i.e., no spiral arms) that are round or slightly elongated (i.e., not flattened into a disk). The stars move every which way: some go around the center as in spiral galaxies, but others plunge toward the middle, come out the other side, then turn around and repeat the process. In principle, the rich array of motions offers valuable information about mass. In practice, the analysis is again complicated by the fact that we can measure only one component of the motion. Studies of stellar motions in elliptical galaxies generally support the dark matter hypothesis, although the evidence and interpretation are sometimes ambiguous.

Gravitational lensing puts a new spin on these two methods. First, it uses light rays rather than stars or star clusters to probe a galaxy's gravitational field. For a distant galaxy, the amount of bending turns out to be comparable to the galaxy's angular size, so lensed images appear a little outside the visible extent of the galaxy. That is a prime region for studying dark matter. Second, the majority of lens galaxies are ellipticals because those tend to be more massive than spirals and hence more likely to produce gravitational lensing effects.

In a sense, lensing provides both more and less information than stellar motions. With lensing, we can measure the side-to-side motion—that is the bending itself. We cannot necessarily measure the absolute amount of bending, because the exact position of the background source is often unknown, but we can measure the relative bending of multiple images since they must come from the same source. As the light rays pierce the galaxy's halo from back to front, they are sensitive to all of the mass along the line of sight. Ultimately, gravitational lensing involves different physical processes, observational approaches, and analysis methods than stellar motion studies, so it provides a valuable complement.

There are some limitations, however. We do not get to choose which galaxies to study with lensing; we can only use the ones that happen to have a bright source in the background. Even when we do see lensing, the images probe the gravitational field at a limited set of locations around the galaxy. Trying to determine a galaxy's mass distribution by examining four lensed images, for example, is a little like trying to figure out what is on a table by placing a scale under each leg. We can get a pretty good sense of the total mass on the table, and we can deduce whether it is spread out or concentrated on one side, but we cannot tell whether the mass is the collected works of Shakespeare or a watermelon.

What can we learn from lensed images? Suppose we see a perfect, round Einstein ring. We immediately know that the

3.2 Compared to a low-mass lens (left), a high-mass lens (right) creates more bending and therefore produces a larger Einstein ring.

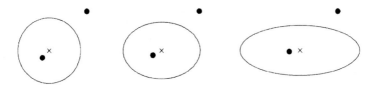

3.3 The left panel shows that a circular lens produces two images (circles) on opposite sides of the lens center (cross). The middle and right panels show that elongated lenses can produce two images that are not aligned with the center.

galaxy is round (otherwise it could not produce a perfect ring). We also know that the source is directly behind the lens. The size of the ring therefore reveals the amount of light bending and hence the mass of the galaxy (strictly speaking, the mass enclosed by the ring; see figure 3.2). Suppose instead that we see two images. They will be lined up with the center of the galaxy if it is round, but usually offset if it is elongated (see figure 3.3). The distances of the images from the center still tell us about the mass of the galaxy (even if the details are a little more complicated than for the ring). If there are four images, the galaxy must be elongated, and the image positions again let us measure the galaxy's mass.

While those are the key principles, the actual analysis is more quantitative. It involves building models for the mass distributions of lens galaxies. In scientific usage, a model is a simplified representation of a system that captures the key aspects and omits complicating details. A basic lens model usually includes an ellipsoidal mass distribution representing the main lens galaxy, plus a background source that may have just a position and brightness (for a quasar lens) or may be more complicated (for an Einstein ring). Varying the parameters of the galaxy and source reveals the range of models that can reproduce the lens data, which in turn determines how specific or uncertain our conclusions will be. The analysis may get fairly complex, especially if we go beyond basic models, but still the first things we usually learn about a lens galaxy are its mass and shape.

What do those quantities teach us? First, lens galaxies need dark matter. The mass inferred from lensing is typically larger than the mass associated with the stars. Second, the dark matter is more spread out than the visible matter. This conclusion comes from comparing lensing (which probes mass at larger radii, where the images are) with stellar motions (which generally probe mass at smaller radii, where most of the stars are). It also comes from comparing different lens galaxies to each other in a statistical analysis. Together, all of the evidence suggests that the mass distributions in elliptical lens galaxies are quite similar to the mass distributions that give rise to flat rotation curves in spiral galaxies. Both types of galaxies require dark matter halos that are much more extended than the distribution of stars.

The commonality between ellipticals and spirals is striking for several reasons. The mix of stars, gas, and dark matter varies throughout a galaxy, yet the rotation speed remains (nearly) constant. Somehow the luminous and dark matter arrange themselves as a galaxy forms to produce a flat rotation curve, even though they can "talk" to each other only through gravity. And this seems to happen in both spiral and elliptical galaxies, even though they have very different formation processes. A spiral galaxy forms when dark matter is drawn together by gravity into a halo, and the luminous matter cools and settles into a disk within the halo.[3] An elliptical galaxy forms when two spirals crash together; the collision rearranges the stars and dark matter.

The conditions that allow a flat rotation curve can vary substantially from one galaxy to the next, and yet most galaxies seem to find them. Surely this is telling us something about dark matter and galaxy formation, but the message has not been decoded yet. A third lesson involves shapes. On the whole, lensing indicates that galaxy mass distributions are round or moderately elongated; they are certainly not as flattened as the disks of spiral galaxies. Elliptical galaxies themselves are not highly flattened, so the shapes inferred from lensing are not particularly surprising. Still, it is important to verify that dark matter halos can be relatively round, and their shapes need not be the same as the shapes of the galaxies they contain.

Studies of lens galaxy shapes also reveal that a galaxy's environment is important; many galaxies have neighbors that affect the light bending. To understand how, think of the gravity between Earth and the Moon. As discussed in chapter 1, Earth's gravity exerts a force that keeps the Moon in orbit. By Newton's third law of motion (equal and opposite reactions), the Moon exerts a force that causes Earth to move a little. We often say that the Moon orbits Earth, but strictly speaking they both orbit the center of mass (which is inside Earth but not right in the middle). That is not all, however. Because gravity weakens with distance, the Moon pulls a little stronger on the side of Earth facing the Moon, and a little weaker on the side facing away. This has little effect on rock, which is stiff, but more effect on water. On the side of Earth facing the Moon, the stronger gravity pulls water up and away from Earth's surface, creating a bulge in the oceans. As Earth's surface rotates through the bulge, we experience high tide. What happens on the side of Earth away from the Moon is a little more subtle, but you can picture the Moon trying to pull Earth out from under the oceans, which also winds up raising the water relative to Earth's surface. In other words, the Moon's gravity stretches Earth's oceans to create two tidal bulges—which is why there are two cycles of high and low tides every day.

A galaxy's neighbors exert a similar **tidal force** that influences how the galaxy bends light. This effect might seem like a nuisance in lens models, because it forces us to identify neighbors and account for their gravity if we want to reproduce the high-quality data available today. Yet it provides an opportunity to study the connection between a galaxy and its neighborhood. We can use lensing data to identify galaxies whose environments are important and target those for further study. Examining lenses and their environments together offers a two-pronged approach that links gravitational lensing to the larger cosmological endeavor of understanding how galaxies grow and evolve.

To recap, the general lesson from stellar motions and gravitational lensing is that galaxies contain a significant amount of matter that we do not see, which is distributed in halos that are relatively round and much larger than the galaxies themselves. When we look at galaxies, the part we can see is like the tip of an iceberg. There is much more to find if we can reach beneath the surface.

CHAPTER 4

Is Dark Matter MACHO or WIMPy?

WHAT IS THE SUBSTANCE that enshrouds galaxies? Our first guess might be something familiar that is just hard to see. We can certainly imagine that dim things could be out there. Some of the brightest objects in our night sky (such as the Moon, Venus, and Jupiter) are vivid only because they are illuminated by a star. If these bodies were moved out of the solar system, they would become extremely hard to see. Amassing enough planets might be challenging; even giant Jupiter is 1,000 times less massive than the Sun, so it would take roughly 1,000 times more planets than stars to have as much "dark" mass as stellar mass. Yet there is a lot of space between stars, and it could be pretty well filled with planets without substantially changing what we see in the night sky.

Another possibility is "failed" stars known as **brown dwarfs**. With masses between about thirteen and eighty times the mass of Jupiter, these objects are too small to shine as normal stars do. In a star like the Sun, gravity squeezes the gas to create immense pressures and temperatures and initiate nuclear reactions that transform hydrogen into helium. This **nuclear fusion** converts mass into energy according to Einstein's famous equation $E = mc^2$, and the energy works its way to the surface to be released as light. Below about eighty Jupiter masses, however, the temperature is too low to ignite hydrogen fusion, so the ball of gas remains largely inert. If brown dwarfs filled the galaxy, they would not be much more visible than lone planets.

A third possibility is stellar corpses. In some five billion years, the Sun will run out of fuel, and fusion will cease to produce energy. Gravity will then crush the gas until the electrons push back, at which point the Sun's mass will be packed into a dense object known as a **white dwarf** that is about the size of Earth.[1] White dwarfs are hot enough to glow blue-white, but small enough to be much less luminous than normal stars. In stars more than about eight times the mass of the Sun, gravity is strong enough to overwhelm the electrons, squeezing them together with the protons to turn all of the mass into neutrons. A **neutron star** has a little more mass than the Sun crammed into the volume of a city (the radius of a neutron star is only a few kilometers). In stars that are still more massive, gravity conquers even the neutrons and causes all of the mass to collapse into a **black hole**. In all of these cases, the matter exists in extraordinary states that do not occur on Earth, but it is still fundamentally normal matter made of the same types of particles as we are.

Dark matter might, however, be something entirely different. Before discussing exotic possibilities, it is important to review the particles known to physicists. The **Standard Model of particle physics** (see figure 4.1) contains twelve fundamental constituents of matter.[2] The **quarks** in the top two rows have been given whimsical names that belie their importance as the components of key particles: two up quarks and one down form a proton, while two downs and one up form a neutron. Joining the up/down quarks and electrons in the first column are neutral particles called **neutrinos**. These low-mass, fast-moving particles are associated with reactions that create or destroy electrons (such as fusion in the Sun; see below).

Together, up quarks, down quarks, and electrons comprise all of the elements in the periodic table; everything in chemistry and biology, in other words, emerges from just the first column

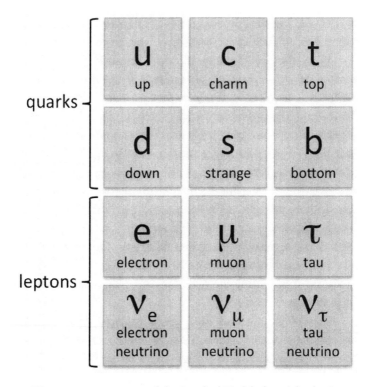

4.1 The matter components of the Standard Model of particle physics. The top two rows contain particles known as quarks that come together in pairs or triplets to form other particles; for example, a proton contains two up quarks and one down quark. The bottom two rows contain particles that can be found individually and are known generically as **leptons**.

of the Standard Model. What about the other columns? The **muon** and the **tau particle** are similar to the electron but more massive. Muons are seen in particle showers produced when cosmic rays hit Earth's atmosphere (and they played a key role in testing Einstein's theory of relativity in the 1960s). The other four quarks create a panoply of particles seen at accelerators like the Large Hadron Collider. In general, particles that emerge from the second and third columns of the Standard Model are short-lived, but they have identifiable signatures such that all twelve fundamental constituents have been discovered.

All of the particles have alternate forms known as **antimatter**. An anti-electron, for example, is identical to an electron except that its electric charge is positive rather than negative. Antimatter participates in particle reactions according to rules about what can and cannot change. For example, fusion in the Sun involves four hydrogen nuclei combining to form one helium nucleus. The four original hydrogens have a combined electric charge of +4, but the final helium has a charge of +2 (it contains two protons and two neutrons). The total charge has to remain the same, so the other +2 must go into two anti-electrons. That, in turn, requires two neutrinos. The particle rules may seem a bit complicated, but they have been extraordinarily successful at predicting the outcomes of experiments in particle accelerators.

Theoretical physicists have long speculated about particles beyond the Standard Model. To some extent, these might just be theorists' dreams, results of late-night "what if" sessions. But precedent suggests otherwise. Around 1930, the anti-electron was predicted by Paul Dirac and then discovered by Carl Anderson. Apparent gaps in the Standard Model led to predictions and subsequent discoveries of the top quark (in 1995) and the tau neutrino (in 2000). The **Higgs boson** was predicted by several physicists in the 1960s before being discovered at the Large Hadron Collider in 2012–13. Theorists can make a strong case that their predictions should be taken seriously. (The Nobel Prize committee concurred, awarding the 2013 Nobel Prize in Physics to Peter Higgs and François Englert for their prediction of the particle that became known as the Higgs boson.)

How does this relate to dark matter? While known particles do not have the right properties to be dark matter—they are charged, or short-lived, or in some other way unsuitable—some of the speculative particles might fit the bill. One popular class of candidates is known as **Weakly Interacting Massive Particles,** or **WIMPs,** because they experience only two of the four fundamental forces in physics: gravity (because they have mass) and the weak nuclear force (which is associated with radioactivity). They do not feel either the electromagnetic force (because they are electrically neutral) or the strong nuclear force (which holds atomic nuclei together). WIMPs are considered good candidates for dark matter because weak force interactions immediately after the big bang[3] would have produced an abundance of particles that is surprisingly similar to what we infer for dark matter. Many people think that the quantitative agreement between cosmology and particle physics is not a coincidence but a suggestion that particles interacting via the weak force may be just what the cosmologists ordered.

As these ideas were developed in the 1980s, it became clear that there were two opposing interpretations of dark matter. In the particle physics corner were the WIMPs. In the astrophysics corner were the objects that Kim Griest thought were just begging to be called **MACHOs**—for **Massive Astrophysical Compact Halo Objects.**[4] Dark cosmology became, for a time, a battle between MACHOs and WIMPs.

How could we tell the difference? One point is that MACHOs may be dim but not entirely dark. Most astrophysical objects emit at least a little light, depending on their temperature. White dwarfs have surfaces that are actually hotter and brighter than the Sun's, but they have much less surface area and wind up being less luminous. Brown dwarfs and planets are relatively cool, so their emissions are dim and mostly in the infrared portion of the electromagnetic spectrum (similar to the light that thermal imaging cameras record). Neutron stars often emit rotating beams of radio waves that we detect as periodic pulses. Black holes themselves do not produce light, but matter falling into a black hole can be heated to enormous temperatures and emit X-rays before crossing the point of no return.

That said, many MACHO candidates would be hard to see directly unless they are relatively nearby. It would therefore be better to design a search based on mass rather than light. Here the key point is that if dark matter is MACHOs then a lot of mass is bound up in discrete objects, whereas if dark matter is WIMPs then the extra mass is spread in a smooth sea within and around galaxies. This difference involves the physics of star formation. Gas must cool and condense in order to go from diffuse primordial clouds to dense objects such as stars and planets, and the best way to do this is to emit light. (Heat and light are both forms of energy; in hot gas, heat energy can be converted into light that carries the energy away.) Normal matter can do that, but WIMPs cannot; thus, there can be no such thing as a WIMP star.

The distinction turns out to be important for gravitational lensing. In 1936, Einstein dismissed the phenomenon as too unlikely to be interesting. Fifty years later, Bohdan Paczyński realized that MACHO dark matter would enhance the likelihood of lensing. He also suggested a promising method to look for the phenomenon. If a star or MACHO moves in front of a more distant star, lensing will focus light and make the background star appear brighter than normal. Strictly speaking, lensing creates multiple images, but they are too close together to be resolved, so we just look for the overall change in brightness. As the stars move apart, the background star will return to its natural brightness. Given how fast stars move, one of these **microlensing** events in our galaxy would last somewhere between a few weeks and a few months.

Even with a boost from MACHOs, microlensing would still be rare; astronomers would have to watch millions of stars to see a handful of them get lensed. But that was becoming conceivable with the advent of digital imaging and computer processing. And with the potential to identify dark matter, the effort began to seem worthwhile. Several teams set out to search for microlensing, including the MACHO Project,[5] the Optical Gravitational Lensing Experiment (OGLE), and Expérience pour la Recherche d'Objets Sombres (EROS).

Their plan was clever. Before they could use microlensing to search for MACHOs, the teams had to show that the phenomenon does occur and can be detected. They set out to observe the center of the Milky Way because looking straight through the main population of stars in our galaxy would increase the chance of seeing microlensing. An event when one star lenses another would not say anything about dark matter, but it would demonstrate that the observations and analysis methods are effective. The teams also needed identify brightness changes caused specifically by microlensing. They analyzed both blue and red light in order to separate microlensing (light bending does not depend on color) from natural variability (stars that change brightness usually change color as well). Finally, the teams had to find places where they could look for source stars that are behind a lot of dark matter. They looked toward a small galaxy orbiting the Milky Way known as the Large Magellanic Cloud (LMC). Light from the LMC travels about 160,000 light years through the dark matter halo of the Milky Way on its way to Earth.

The effort was painstaking but ultimately worthwhile. In a three-year campaign, the MACHO Project saw ninety-nine events among seventeen million source stars in the Milky Way. At the most basic level, the results demonstrated that microlensing occurs as predicted and that the events can be extracted from the reams of observational data produced by the survey. (They also provided new information about the distribution of stars in the Milky Way, but that is not part of our main story.) For the dark matter investigation, the MACHO Project saw a little over a dozen events in six years among twelve million stars in the Large Magellanic Cloud. (The official number was between thirteen and seventeen, depending on the criteria used to classify events.) The other teams obtained similar results. The analysis was rather involved because microlensing can be caused not only by MACHOs but also by stars in the disk of the Milky Way, in a stellar halo around the Milky Way, or in the LMC itself. In the end, the teams determined that the number of events was smaller than the number expected if the Milky Way's dark matter were all MACHOs. Dark matter, in other words, is not just normal matter in a form that is hard to see.

This conclusion is consistent with another line of inquiry involving the early universe. During the first few minutes after the big bang, nuclear fusion turned about 25 percent of the hydrogen into helium and produced trace amounts of other elements such as deuterium and lithium. Countless generations of stars have polluted much of the universe's gas in the intervening 13.8 billion years, but it is possible to find pockets of gas that is still relatively pristine. Measurements of the abundance of light elements yield upper limits on the density of normal matter in the universe, which also indicate that the bulk of the matter in the universe is not something familiar.

By process of elimination, we infer that dark matter must be some exotic particle. Does that mean it is a WIMP? Not necessarily; there are other possibilities. But a WIMP is certainly a good candidate worth exploring further. That leads to the question, Can we search for WIMPs, and if so, how?

The answer is yes, and it has two aspects. First, as discussed in chapter 1, if dark matter is WIMPs then particles are passing through us all the time. From the perspective of the Milky Way, the Sun and Earth are flying through a sea of dark matter at hundreds of thousands of miles per hour; from our perspective on Earth, we experience a "WIMP wind" blowing at that incredible speed. Yet we hardly notice the particles, and they hardly notice us; a WIMP could travel through a solar system's worth of lead with no more difficulty than it travels through empty space. The second aspect involves that word "hardly," which in turn relates to the first two letters in the acronym: weakly interacting. If dark matter does feel the weak nuclear force, then it can interact with the nuclei of normal atoms even if the interactions are, well, weak. Here "weak" means that the interaction rate is low; the probability that a given dark matter particle will collide with a piece of Earth is very small. But the flow of particles is high enough that collisions should occur every once in a while. This creates the opportunity to catch dark matter here on Earth.

While this may sound abstract, physicists already know that we are bathed in a shower of ghostly particles. As mentioned earlier, nuclear fusion in the Sun produces a huge number of elusive particles called neutrinos. More than ten billion billion neutrinos from the Sun fly through your body each day, but only a few actually hit you (and even those have little effect). The story of neutrinos is a bit of a sidetrack, but one that is instructive.

In the 1960s, physicists and astronomers got interested in using neutrinos to see how well we understand the conditions at the center of the Sun. Raymond Davis and John Bahcall realized that neutrinos interact more readily with a particular isotope of chlorine than with most atoms, and the reactions can be identified because they convert chlorine atoms into argon atoms that can be extracted chemically. The interactions remain rare, so it takes a lot of chlorine to capture a detectable number of neutrinos. Fortunately, chlorine atoms are easy to come by; they are part of a common cleaning fluid called tetrachlorethylene. Davis filled a tank with 100,000 gallons (600 tons) of cleaning fluid, connected a device that could pull out argon, and began searching for neutrinos. He put the equipment almost a mile underground, in an abandoned South Dakota gold mine, to protect it from a steady "background" of particles at Earth's surface. (For example, particles from natural radioactivity or cosmic rays could mimic neutrinos by interacting with chlorine to produce argon.) The mile of rock above the mine blocks most background particles, but not neutrinos. Today, many experiments designed to detect neutrinos or dark matter follow Davis's lead and operate underground. One called IceCube even operates under the surface of Antarctica, using the ice as both shield and detector.

Davis's experiment did see neutrinos (leading to a share of the 2002 Nobel Prize in Physics), but not nearly as many as expected. The disagreement—dubbed the "solar neutrino problem"—led people to rethink both our models of the Sun and our understanding of neutrinos. Astrophysicists had a hard time getting the models to predict many fewer neutrinos. Particle physicists realized, however, that neutrinos might be even more slippery than people thought. Theorists were beginning to wonder if an electron neutrino from column 1 in the Standard Model could spontaneously change into a muon neutrino from column 2 or a tau neutrino from column 3. The idea was important because Davis's chlorine tank could catch only electron neutrinos. If electron neutrinos produced in the Sun turned into muon on tau neutrinos before they reached Earth, they would not have been detected by an experiment like Davis's.

The operative word was "if." Theorists understood that neutrinos can change "flavor" only if they have some mass, but no one knew whether that was the case. Astronomers and particle physicists together realized that building new detectors sensitive to muon and tau neutrinos in addition to the electron variety might simultaneously solve the solar neutrino problem and answer long-standing questions about these puzzling little particles.

In Japan, an experiment called Super-Kamiokande used a tank containing 50,000 tons of water. A neutrino moving fast through water produces distinctive light known as Čerenkov radiation. Such an experiment can record all three flavors of neutrinos but cannot tell them apart. In Canada, the Sudbury Neutrino Observatory (SNO) also used water, but a special kind in which the standard hydrogen is replaced by a variant called deuterium. Such "heavy water" is relatively rare—SNO could obtain only about 1,000 tons—but it offers different types of interactions that make it possible not only to detect but also to distinguish the three flavors of neutrinos. Together, Super-Kamiokande and SNO revealed that neutrinos do in fact change flavors, proved that neutrinos have mass, and demonstrated that the total number of neutrinos reaching Earth does match expectations.

The neutrino saga is not directly related to dark matter; neutrinos themselves do not have enough mass and they move way too fast to be the "cold" dark matter needed to explain galaxies and other cosmic structures. But the saga does have some important indirect lessons. At a conceptual level, it demonstrates how much we can learn when astrophysicists and particle physicists work together, attacking a problem from opposite directions. At a practical level, it provides experience with designing and running experiments to catch fleeting particles from space.

A number of groups have taken the lessons to heart and built WIMP detectors at underground sites around the world. Various phenomena can register a WIMP, but three are most widely used. First, a WIMP passing through translucent material can create a flash of light through *scintillation*. Second, a WIMP can disrupt atomic structure and create positively and negatively charged particles through *ionization*. Third, if a WIMP hits a nucleus in a crystal, it can create a *vibration* in the crystal lattice. Different experiments use different combinations of these effects to record the passage of a WIMP through the detector.

Several experiments have claimed to find evidence for WIMPs, but as of early 2014 the results are considered to be inconclusive or controversial. The first claims came from an experiment called DAMA and its successor DAMA/LIBRA. DAMA differs from other experiments by eschewing any attempt to distinguish WIMP detections from non-WIMP background events. Instead, DAMA relies on the fact that the Earth's motion around the Sun changes the speed at which our planet flies through the WIMP sea. In June, our motion around the Sun is in the same direction as the Sun's motion around the Milky Way, so the WIMP wind is a little stronger than average. In December, Earth moves "backward" relative to the Sun's motion, so the wind is a little weaker. DAMA records an event rate that varies throughout the year, peaking in June and reaching a minimum in December, which the DAMA team interprets as clear evidence that they are detecting WIMPs.

Other teams remain skeptical, citing two principal arguments. First, the types of WIMPs that could create the DAMA signal should have been seen in some other experiments but were not. There are ways to arrange for dark matter to be visible to DAMA and invisible to other experiments, but they seem rather contrived. Second, there are many things on Earth that vary in a yearly cycle and might create a modulating background that could be misinterpreted as evidence for WIMPs. The DAMA team claims to have considered this possibility and ruled it out, but skeptics insist that they need more information about the raw data and analysis methods (which the DAMA team will not release) before they can be convinced.

The stalemate can probably be broken only with more data collected independently. Fortunately, many other experiments are under way. An experiment called CDMS, which uses very careful background rejection, recently reported three events that were consistent with WIMP interactions when less than one event from backgrounds would have been expected. Other experiments called CoGeNT and CRESST have also reported signals. However, the putative detections are not all consistent with one another. Furthermore, the two most sensitive experiments to date (called XENON100 and LUX) have not seen any events above the expected backgrounds.

Clearly there is no definitive answer yet, but such is the nature of research that pushes the very limit of our technology and understanding. The experiments and analyses (and bickering) will continue for some years to come, illustrating all the while how science operates at the frontier. The key point for our story is that a lot of physicists are spending considerable time, effort, and money to search for particles that astronomers say must exist. Whether those particles are in fact WIMPs remains to be seen; they could turn out to be something altogether different. What we do know is that MACHOs have been eliminated from the fight, and gravitational microlensing helped deliver the knockout blow.

CHAPTER 5

Finding What's Missing

WHILE THE QUEST TO UNDERSTAND dark matter
has been taken up by physicists, it still has valuable contributions
from astrophysics in general and gravitational lensing in particular.
Here the story shifts back to lensing by distant galaxies, and it has
two threads that merged at the turn of the twenty-first century.

The first thread involves lens systems in which a foreground
galaxy creates four images of a background quasar. In the 1990s,
we began to gather a sizable sample of four-image systems and
observe them with the Hubble Space Telescope and radio telescopes. The sharp pictures let us measure the positions of lensed
images very precisely, with measurement uncertainties that are
300 to 500 times smaller than the data values. Gravitational lens
models that account for the mass, shape, and environment of the
lens galaxy (see chapter 3) can usually reproduce the *positions*
of the images quite well, but not the *brightnesses*. In figure 5.1,
one panel shows observations of a four-image gravitational lens
system made with a radio telescope, while the other panel shows
predictions from a lens model. The model can put images in the
right places, but it predicts that the two images labeled A and
B should be nearly equal in brightness when that is clearly not
the case. Something must be missing from the model, but what?

One thing often omitted from lens models is stars. The stars
in a galaxy create a certain graininess in the gravitational field
that leads to small-scale variations in the lensing magnification.
This is another form of microlensing, and its scale is small indeed:

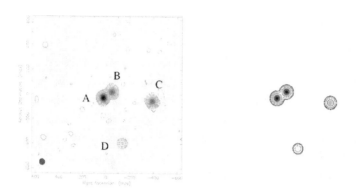

5.1 The left panel shows the four-image lens system B1555+375 observed with a radio telescope. The quasar images are labeled A–D. The lens galaxy is not seen at these wavelengths. The right panel shows predictions from a lens model for B1555+375. The model can reproduce the positions of all of the images but not the relative brightnesses of images A and B. Credit (left): Marlow et al. (1999). © AAS. Reproduced with permission. Credit (right): the author.

the distance over which the magnification changes due to stars is about a million times smaller than the size of the galaxy. Yet quasars are compact enough that their light beams feel the graininess, leading to magnifications that can be quite different from what is predicted by models in which the mass distribution is smooth. We know that the graininess is there because the magnification varies over the course of months and years as the stars and quasar move relative to one another. A number of researchers use this "cosmological microlensing" (as compared with the "galactic microlensing" discussed in chapter 4) to study the populations of stars in lens galaxies and the structure of the source quasars.

Microlensing cannot explain all of the anomalous image brightnesses, however. Recall that figure 5.1 shows data from a radio telescope. The radio light from a quasar is emitted from a much larger region than the optical light—large enough, in fact, that the radio light does not feel the graininess caused by stars. If not microlensing, what can cause radio lenses to be anomalous?

In 1998, Shude Mao and Peter Schneider suggested an answer. They pointed out that lens galaxies could contain intermediate-scale structures that are smaller than a whole galaxy but larger than a single star. For example, globular clusters are collections of hundreds of thousands of stars packed into a region only a few dozen light years across (for comparison, there are only twenty-three stars within a dozen light years of the Sun). A globular cluster would affect gravitational lensing on a scale about a thousand times larger than that of a star but a thousand times smaller than that of a galaxy, through a phenomenon that has been dubbed **millilensing**. In principle, this is just the right scale to create radio lens anomalies. In practice, Mao and Schneider estimated that most galaxies do not contain enough globular clusters to explain why anomalies are as common as they are.

Around the same time, computer simulations of galaxy formation were beginning to give robust predictions about the distribution of dark matter on small scales. A galaxy forms when a kernel of matter whose density is higher than average begins to draw in material from surrounding regions, growing ever more massive and coming to dominate its neighborhood. We know how to describe the process mathematically—it "just" involves equations of motion and gravity—but the equations are far too complicated to be solved exactly. There is, however, a straightforward algorithm for obtaining an approximate solution. If we know the positions and velocities of all the particles at some time t_1, we can use the equations of motion and gravity to calculate how the positions and velocities will change when we take a small step forward to time t_2. We can then repeat the process to advance to time t_3, do it yet again to get to time t_4, and so forth and so on. The algorithm is effective but mind-numbingly tedious—bad for people, in other words, but perfect for computers.

Even computers can handle only a finite amount of complexity, but advancing technology has allowed simulations of galaxy formation to become increasingly detailed. By the late 1990s, researchers began to notice that up to 10 percent of the mass in a simulated galaxy is locked up in small but coherent **subhalos** that orbit within the larger dark matter halo. These subhalos are remnants of a "bottom-up" formation process. When the universe was young, the first objects to collapse under their own weight were small halos of dark matter. As time passed, the small halos collided and merged to form bigger halos, which continued to absorb even more objects from their surroundings. When a small halo fell into a big one, the outer parts were stripped off by the collision to join the common halo enveloping the whole system, but the core was dense enough to remain intact and become a new subhalo. Over the course of billions of years, this process occurred thousands of times to produce dark matter distributions like that shown in figure 5.2. At least, this is what simulations predict.

At first glance, the predictions look good because we know that galaxies are surrounded by small structures. Our own galaxy is orbited by a few dozen dwarf galaxies (including the Large and Small Magellanic Clouds) that are bound by their self-gravity even as they move around the Milky Way. Other galaxies have their own contingents of dwarf satellites. The problems arise when we start counting. In 1999, two research groups led by Anatoly Klypin and Ben Moore pointed out that simulations predict hundreds or thousands of dwarf companions for a galaxy like the Milky Way, whereas only a few dozen are seen. The discrepancy became known as the "missing satellites problem."

What does it mean? One possibility is that our ideas about dark matter are wrong. Changing the physical properties of dark matter could affect the distribution of small-scale structure. For example, if dark matter were *warm* instead of cold, the particles would be able to fly away from low-mass subhalos, preventing small structures from forming in the first place. (Large structures could still form because they have enough gravity to collect even

5.2 Distribution of dark matter in a simulated galaxy. The main halo of dark matter is full of thousands upon thousands of clumpy "subhalos." On this scale, the visible galaxy would occupy a small region in the center of the image. Credit: J. Diemand et al. (2008).

warm particles.) Alternatively, if dark matter were *self-interacting*, two dark matter particles could annihilate and turn into normal particles or light. That would happen mainly in regions where the density is particularly high, so it could wind up erasing small substructures. If either of these hypotheses is correct, the number of dark matter subhalos could be far less than initially predicted, and closer to the number of observed satellite galaxies.

There is another possibility, however. The missing satellites problem is actually a case of comparing apples and crabapples:

the predictions involve subhalos filled with dark matter, while the observations involve dwarf galaxies that are visible because they contain stars and/or gas. As far as we can tell, every galaxy forms within a halo of dark matter. But does it necessarily follow that every halo (and subhalo) of dark matter contains a visible galaxy? The answer appears to be "no." People who study galaxy formation have identified a variety of mechanisms that could prevent low-mass subhalos from hosting any matter that can be seen with our telescopes. If these mechanisms work as imagined, there may be an enormous number of subhalos lurking beyond our vision, never to be seen directly.

The missing satellites problem, in other words, has two possible solutions. In one, the "fault" lies with theorists who have misunderstood dark matter and overpredicted the amount of substructure. In the other, the "fault" lies with observers who have failed to find the vast array of substructure all around us. In bestowing the moniker "missing satellites problem," Klypin and his collaborators revealed their persuasion. Some observers have taken umbrage and suggested that, to be fair, we should at least allow for the option of renaming it the "satellite overprediction problem."

To see who is right, we need to find a way to search for subhalos that contain mass but no light—**dark dwarfs**, if you will. This is where the two threads come together. In 2001, Ben Metcalf and Piero Madau, along with Masahi Chiba working independently, suggested that dark dwarfs might constitute the type of substructure that Mao and Schneider needed to explain anomalous lens brightnesses. The next year, Neal Dalal and Chris Kochanek worked out the numbers, asking how much substructure would be required to explain the observed anomalies. The answer appeared to be consistent with dark matter predictions. Gravitational lensing had, it seemed, found the missing satellites.

As is often the case in science, a great discovery signals not the end of the story but the beginning. The work by Dalal and

Kochanek raised as many questions as it answered. Plus, the prospect of counting dark dwarfs to learn about the fundamental nature of dark matter made it worthwhile to dig into the details.

Wyn Evans and Hans Witt asked the first hard question: Are we thinking about lens models in the right way? More to the point, are mass clumps the only way to explain lens anomalies? Evans and Witt pointed out that standard lens models assume that the mass distribution has a perfect elliptical shape, but real galaxies often have small deviations from ellipses. They suggested that allowing a more general shape might make it possible to explain anomalies without mass clumps. To test this suggestion, Art Congdon and I tried fitting several of the anomalous lenses using generalized mass distributions. The models we found were quite striking; some looked more like Pac-Man than a galaxy. Other studies found that additional data (such as Einstein rings) constrain deviations from ellipses to be very small. Together, these investigations indicate that galaxy shapes alone do not provide a plausible explanation for anomalous lenses. Does this mean that Evans and Witt failed? Certainly not. While their hypothesis was ultimately ruled out, it did stimulate several new research projects that helped us learn more about analyzing and interpreting lens anomalies. In the world of scientific ideas, being tested and eliminated is far better than being ignored.

Another argument that anomalies are real and significant comes from a connection between gravitational lensing and mathematics. The equation that describes lensing turns out to be very rich from a mathematical standpoint. Arlie Petters, Scott Gaudi, and I used a formal analysis of the lens equation to unveil mathematical relations between the brightnesses of lensed images that are *universal* for smooth mass distributions. Those relations can be violated only if the lens has some complicated structure that is smaller than the distance between images. When we applied the relations to close pairs and triplets of images in observed lenses, we identified a lot of brightness anomalies.

Some of them might be caused by stellar microlensing, but others appear to require dark matter millilensing.

The next question is how much we can learn about dark matter substructure from different aspects of lensing. Dalal and Kochanek used a statistical analysis of brightness anomalies to constrain the overall amount of substructure in lens galaxies. More recent studies have use detailed studies of individual lenses to identify specific clumps. Some people still use the positions and brightnesses of images in quasar lenses, while others have begun to use small distortions in Einstein ring systems. The two types of lenses are fairly complementary in terms of data: quasar lenses offer precise positions and brightnesses, but only for two or four images; ring lenses offer images that are much more extended, but the structure of the source is unknown so the modeling is more complicated. At this point, quasar and ring lenses have both provided good constraints on mass clumps in distant galaxies.

The clumps found so far tend to be on the massive end of the range for substructure. In fact, some of them actually contain visible dwarf galaxies. Such clumps are clearly not *dark* dwarfs, but they are interesting nonetheless. For one thing, they confirm that mass clumps can be reliably inferred from lens models. Also, dwarf galaxies detected in this way can be compared with dwarf galaxies near the Milky Way to learn more about this type of galaxy. Finally, the massive clumps that have been identified are presumably the tip of the iceberg in terms of substructure. Otherwise it would be unlikely to find subhalos that lie close enough to the images to create strong perturbations.

Stating that claim is quite different, of course, from proving it. The next goal is to find direct evidence for a larger population of subhalos. This is where the best opportunity to learn about dark matter lies. Recall that conventional cold dark matter predicts substructure that extends down to low masses, whereas alternatives like warm or self-interacting dark matter can yield many fewer low-mass subhalos. Counting clumps as a function of mass would test the different hypotheses. How can we do it? Theoretical studies suggest that clumps near images may not be the only ones that matter; clumps far from the images might have small effects individually but significant effects collectively. For example, Leonidas Moustakas and I showed that lens time delays feel the accumulated effects of mass clumps throughout a lens galaxy, providing a way to probe substructure that complements what we can do with brightness anomalies.

The next decade will bring more opportunities to shed light on dark matter with gravitational lensing. The astronomical community is undertaking a variety of large surveys that will benefit lensing in several ways. Some surveys will yield very sensitive images over as much as half the sky, revealing thousands of new quasar and ring lenses that can be used to probe substructure (among other things). Other surveys will observe the same sources many times to see how they change with time, making it possible to measure lens time delays. A project called the Large Synoptic Survey Telescope (LSST) will combine both approaches, observing all of the sky that is visible from its mountaintop in Chile every few days beginning in 2022. LSST is designed to study all sorts of variable sources, not specifically aiming for lensed quasars but naturally including a huge number of them. The surveys will produce a flood of information, so people are already thinking about how the different types of data will constrain substructure.

To me, what is fascinating about this work is that microscopic properties of dark matter have consequences on astronomical scales, which can be probed with gravitational lensing. Furthermore, this is a field in which observational astronomy, theoretical astrophysics, statistical methods, and formal mathematics all combine to create a rich intellectual framework for studying a substance that may forever elude our grasp. As I said at the outset, when we couple curiosity and creativity to scientific critical thinking, we truly have the potential to discover what's in the dark.

CHAPTER 6

"A Long Time Ago in a Galaxy Far, Far Away"

SO FAR WE HAVE THOUGHT about using lensing the way a doctor uses X-rays, analyzing the way light travels through something to learn what's inside. But we can shift our focus to the light sources themselves and use lensing to help us study objects that would otherwise be too small and/or faint. We can, in other words, use gravitational lenses as cosmic telescopes that boost our Earth-bound telescopes and let us peer further into the depths of the universe.

Our telescopes already act as time machines. The speed of light is large but finite, so it takes a certain amount of time for light to reach us. The effect is hardly noticeable on Earth but significant in space because the distances are so vast. When we look at the Sun, we are seeing light that left the star eight minutes ago. When we view Jupiter, we see the planet somewhere between thirty-five and fifty minutes in the past (depending on the locations of Earth and Jupiter in their orbits around the Sun). We see Neptune as it was about four hours ago. We see the nearest star as it was more than four years ago. And so on. When we look at the nearest big galaxy (called Andromeda), we are looking more than two million years into the past, and with other galaxies we can reach back billions of years.

This trick of light is great for studying the development of the universe. Just as human historians examine the forces that shaped modern societies, cosmohistorians analyze the forces that created the galaxies and clusters of galaxies that populate the universe today. They ask questions such as, Why are there two big classes of galaxies, namely spirals and ellipticals? Why do the different types of galaxies tend to prefer different environments? Did the galaxies form differently, or did they grow different over time? For that matter, when did the first galaxies form? The nice thing for cosmohistorians is that telescope time machines offer the chance to observe the action directly—to see what was happening "a long time ago, in a galaxy far, far away."

With the right equipment we can see back to the big bang itself. Well, almost. For about 400,000 years after the bang, the universe was so hot that no atoms could form; any electron that bound to an atomic nucleus was quickly knocked away by another particle. The sea of charged particles caused light rays to scatter all over the place, making the universe opaque like fog. Only after 400,000 years did the universe cool to the point that electrons and nuclei could combine to form neutral atoms, which are not very effective at scattering light. The light that was liberated then has been traveling for some 13.8 billion years, and it has been stretched by the expansion of the universe to the point that it is now found in the microwave region of the electromagnetic spectrum. This cosmic microwave background (CMB) light was discovered by accident in New Jersey when Arno Penzias and Robert Wilson were testing a new microwave antenna in 1964. It has been mapped in increasing detail by satellites including the Cosmic Background Explorer (COBE), the Wilkinson Microwave Anisotropy Probe (WMAP), and now the Planck mission, along with a variety of specialized telescopes on the ground.

Together optical, infrared, and microwave telescopes are assembling a pictorial history of the lives of galaxies. It's curious that we have great baby pictures (from the CMB) but little knowledge of youth and toddlerhood. While the CMB is everywhere, nascent galaxies were scattered hither and yon. Plus, to see galaxies that far back in time we have to look very far away, which means the objects appear very small and faint. Telescopes have to stare at the same patch of sky for hours or even days on end to pick up tiny smudges of light from the young universe. The effort began in earnest in 1995, when the director of the Space Telescope Science Institute invested about ten days of time on the Hubble Space Telescope to observe the Hubble Deep Field.[1] Similar efforts with new instruments led more recently to the Ultra-Deep Field and then the Extreme Deep Field.

Gravitational lensing aids this endeavor by magnifying distant sources, making them easier to detect and study. Some images are more magnified than others; to quantify that effect, we can use lens theory to predict the magnification as a function of position. Figure 6.1 shows the **magnification map** for the lens from figure 2.8. Lensing magnification can be substantial—a factor of 10 or even 100. Those high magnifications occur only in small regions, however, and randomly distributed sources may or may not lie in the right places.

If we want to take advantage of the lensing boost, we need to find lenses that are both bigger and better. Bigger really means more mass. Clusters of galaxies are the most massive gravitationally bound structures in the universe; some of them contain more than a quadrillion times the mass of the Sun. These are important targets for many different reasons: they are complex environments with a variety of processes that affect galaxy evolution; they are good probes of cosmology because their abundance depends on an interplay between gravitational attraction (driven by dark matter) and cosmic repulsion (driven by dark energy); and they serve as powerful cosmic lenses (recall figure 2.13). Cluster lenses have complex magnification maps that reflect contributions from galaxies, gas, and dark matter (see figure 6.2). For all of these reasons, six cluster lensing fields have been chosen for the next big director's campaign with the Hubble Space Telescope.

0.7   1.3   3.7   13   50

6.1 An example of a lensing magnification map. The color indicates the lensing magnification at each position, using the scale shown at the bottom. The lens is the same as in figure 2.8.

"Bigger" is good, but let's not forget about "better." While mass is certainly important, it is not the only factor that determines whether a lens has good magnifying power. I am part of a team that has asked whether special configurations of mass can enhance the lensing. We have found that the best cosmic telescopes often have multiple clusters that are fairly well aligned from our point of view. Having multiple clusters ensures that

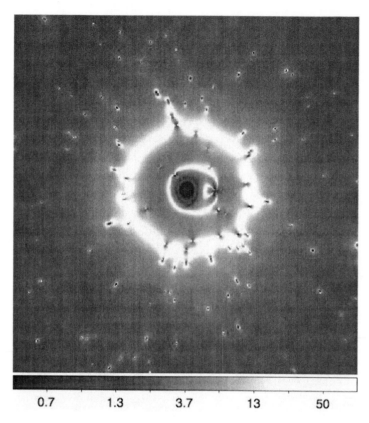

| 0.7 | 1.3 | 3.7 | 13 | 50 |

6.2 Magnification map for a cosmic telescope studied by Ammons et al. (2013). © AAS. Reproduced with permission.

there is a lot of mass in the first place. Clusters that are extremely massive are also extremely rare, so it may actually be easier to find a set of modest clusters than a single cluster with the same total mass. When a cosmic telescope has multiple components, the lensing effects can combine in a way that makes the whole more than the sum of the parts.

It might seem unlikely to find several clusters lined up in space, but the universe is a big place. Today's large computer simulations let us predict that there could be several hundred very good multiple-cluster lenses in the sky. Also, large digital catalogues of the sky (such as the Sloan Digital Sky Survey) provide the data we need to search for outstanding cosmic telescopes. Our team recently identified two hundred of the most promising cases and began to study them in detail.

Whether we use single- or multiple-cluster lenses, we need to think of them as precision tools that couple to our conventional telescopes. Whenever we use a piece of equipment—microscope, telescope, camera, or gravitational lens—to look at something, we need to know how it changes what we see. This is particularly important with lensing because the changes are not uniform: some sources are magnified a lot, others just a little, and we must figure which is which in order to know what the sources really look like.

Calibrating a cosmic telescope basically amounts to mapping the mass. This problem is both easier and harder than mapping matter in galaxies: easier because more data are available, but harder because the mass distribution is more complicated. Let's first consider the data. The galaxies in a cluster are visible, of course, so we know where they are located and how bright they are. Each cluster contains an enormous cloud of gas that actually outweighs all of the galaxies and accounts for most of the normal matter in the system. The immense gravity of a cluster squeezes the gas and heats it to millions of degrees, causing it to glow with X-ray light that can be detected with special telescopes in space (such as the Chandra X-ray Observatory and XMM-Newton). The galaxies and gas are in turn surrounded by a halo of dark matter that may be some ten million light years across. We cannot see the dark matter, naturally, but we can infer its presence from the gravity it exerts on the galaxies and gas, along with gravitational lensing. While it might seem like cheating to use lensing to calibrate a lens, the argument is not actually circular. The ultimate goal is to determine how much small, distant sources are magnified. If we choose our cosmic telescopes well, we will see other sources that are less distant but still far enough behind the cluster to be lensed into small arclets or even giant arcs (see figure 2.13). Many more background galaxies (tens or hundreds of thousands) experience **weak lensing**, or tiny shape distortions that demand sophisticated data processing and statistical analysis yet offer further information about the light bending. We can use lensing of the less distant sources to learn about the mass distribution, then turn around and predict the magnifications of the more distant sources.

What are the complications? We know that the galaxies in a cluster have a significant effect on lensing—they create the bumps and wiggles in the magnification map in figure 6.2—but we do not know how much they weigh individually. We can make a reasonable guess based on relations between brightness and mass determined from stellar motions and gravitational lensing in other galaxies. However, those relations have a lot of "scatter": galaxies with similar brightnesses can have a range of masses. The brightness/mass relations are therefore good on average but not highly precise for individual galaxies. Furthermore, we expect (from both theoretical reasoning and observational evidence) that galaxies have their outer parts stripped away when they fall into a cluster. This is a consequence of the tidal forces mentioned in chapter 3; imagine the Moon not just raising Earth's oceans a little bit but actually lifting the water off the surface altogether. The mass that gets ripped out of galaxies in a cluster is what forms the common halo in the first place. The problem is that we do not know how strong the "tidal stripping" has been, so we do not know in detail how much mass is still attached to the galaxies and how much is in the surrounding halo.

Another challenge lies in describing the distribution of dark matter. Computer simulations make fairly specific predictions about the degree to which dark matter should be concentrated toward the center of the cluster; but nature does not necessarily abide by our predictions! Flippancy aside, there are legitimate concerns because predictions are usually drawn from simulations in which clusters have reached some sort of equilibrium state, whereas any given observed system may have experienced some "recent" events that stirred up the mass. One well-known cosmic telescope is called the Bullet Cluster because a smaller cluster shot through a second, larger one, disrupting the distribution of mass and gas in the system (see figure 6.3). Such major collisions are expected to be rare, but smaller disturbances may not be.

One final complication comes from mass along the line of sight. Even if most of the light bending occurs within a massive cluster, there may be additional small deflections that arise when a light ray passes other galaxies on its way from the source to the cluster or from the cluster to Earth. Theoretical studies suggest that the extra deflections and distortions cannot be ignored, and observational studies support the notion: when line-of-sight structure is neglected, there are small but significant differences between models and data. Many research teams argue that the residual differences do not affect the main scientific conclusions from lens models, but our team is trying to face the issue head-on by measuring line-of-sight structure and incorporating it into our analysis of cosmic telescopes.

The constraints from data help pin down the things we cannot measure directly. For example, weak lensing offers a global view of the mass distribution that may not have exquisite spatial resolution but does not depend on particular model assumptions. A weak lensing analysis of the Bullet Cluster system yields the mass distribution depicted in figure 6.3. Lensing reveals that most of the gravity originates in two concentrations that are centered on the two visible clusters of galaxies. By contrast, most of the light (in the form of X-rays) is found between the clusters. When the two clusters encountered each other, the enormous clouds of hot gas crashed together and slowed down, but the

6.3 Composite image of the Bullet Cluster system. The galaxies are observed at visible wavelengths. The two regions in the middle indicate X-ray light from hot gas. The two outer regions indicate dark matter inferred from gravitational lensing. Image credit (X-ray): NASA/CX-C/M. Markevitch et al. (Optical): NASA/STScI; Magellan/U. Arizona/D. Clowe et al. (2006). (Lensing map): NASA/STScI; ESO WFI; Magellan/U. Arizona/D. Clowe et al. (2006).

galaxies and dark matter kept moving and left the gas behind. This system provides the most striking visual evidence to date that dark matter is fundamentally unlike normal matter. It also illustrates that clusters are not always found in a relaxed, equilibrium state.

That kind of analysis gives a general idea of where the mass is, but to calibrate a cosmic telescope in detail we need to zoom in and probe the mass distribution at higher resolution. Here the other types of data come on stage. For example, strongly lensed arcs and arclets probe the concentration and shape of the dark matter halo on scales too small to be constrained by weak lensing. Also, the brightnesses of the galaxies provide estimates of their masses (as discussed above), which we can use to account for small-scale features in the lensing magnification map. The analysis becomes fairly involved, but a comprehensive approach to both data and models provides a rich understanding of the lensing properties of cosmic telescopes.

Even so, our knowledge of the mass distributions is never perfect; there are too many important quantities that we cannot measure directly, and even the things we can measure have some finite precision. We simply have to acknowledge the uncertainties and deal with them. For example, when we build galaxies into the mass model, we can first use the average relation between brightness and mass, but then we need to ask how much the models change if each galaxy is a little more or less massive than average (in a way that is consistent with the scatter in the relation between brightness and mass). Also, we need to allow the parameters of the dark matter halo to vary within the range allowed by the data. The general idea is to find the full set of models that are consistent with all available data and see how much they differ on predictions for the lensing magnification. Even better is having multiple research teams do the analysis independently to see if they get consistent results.

All of the uncertainties mean that we cannot uniquely determine the magnifying power of a cosmic telescope. But that is okay. It is actually less important to be exactly right about the magnification, and more important not to be wrong. Suppose a distant galaxy is magnified by a factor of ten. If we think the magnification is precisely fourteen, we will misinterpret the intrinsic properties of the source galaxy. However, if the uncertainties imply a reasonable probability for the magnification to be between, say, nine and nineteen, we can factor that into our interpretation. We can use statistical methods to account for the uncertainties and still draw useful scientific conclusions. In a sense, a lot of cutting-edge research (in astrophysics, at least) amounts to dealing with imperfect information in this way.

I hope I have not given the impression that the situation is hopeless. Far from it—it is rather remarkable, I think, how much we can do with tools we cannot control. We do need to be careful in the analysis, realistic in our assessment of the uncertainties, and skeptical (in a good way) about whether all of the important details have been considered. Those are hallmarks of good science anyway. The care and effort are certainly worthwhile, because many of the known objects that are likely to be distant, young galaxies have been seen through cosmic telescopes. Gravitational lensing has already let us reach back some 13.3 billion years to see galaxies that existed when the universe was only about half a billion years old, and it will continue to play a major role as we search for more galaxies from that epoch and even approach the big bang itself.

Glossary

antimatter. Particles that are just like normal matter except that the electric charges are reversed. For example, a normal electron has a negative charge but an anti-electron has a positive charge.

black hole. A region of space where gravity is so strong that nothing can escape. A star that is much more massive than the Sun can collapse into a black hole when it dies.

brown dwarf. A ball of gas that is bigger than a planet but too small to be a normal star powered by hydrogen fusion. Brown dwarfs have masses between thirteen and eighty times the mass of Jupiter.

cluster of galaxies, or **galaxy cluster**. A large collection of galaxies bound by their mutual gravity and embedded in a halo of hot gas and dark matter. The total mass can range from about 10^{14} to more than 10^{15} times the mass of the Sun.

dark dwarf. A subhalo that contains only dark matter (as compared with a **dwarf galaxy**).

dark matter. Material that exerts gravity but has little or no interaction with light. "Cold" dark matter moves at speeds much slower than the speed of light. By contrast, "hot" dark matter moves at speeds very close to the speed of light, while "warm" dark matter moves at intermediate speeds. Dark matter is typically assumed to have little or no interaction with itself or with normal matter, but "self-interacting" dark matter particles can react with each other to produce other types of particles.

Doppler effect. An apparent shift in the frequency or wavelength of sound or light when the source is moving relative to the observer. Light experiences a blueshift when the source and observer are moving toward each other, and a redshift when they are moving apart.

dwarf galaxy. A small galaxy typically found orbiting a larger one. Dwarf galaxies can be seen because they contain stars in addition to dark matter (cf. **dark dwarf**).

Einstein ring. A gravitationally lensed image in the form of a ring, produced by a source directly behind a lens galaxy. Idealized Einstein rings are produced by spherical lenses. Real Einstein rings can be produced by lenses that are slightly elongated, if the source is large.

elliptical galaxy. A galaxy with a smooth distribution of light in a round or slightly elongated shape. The stars move in all different directions.

globular cluster. A collection of hundreds of thousands of stars packed into a region only a few dozen light years across. Typically found orbiting galaxies.

halo. An extended distribution of matter surrounding a galaxy or cluster of galaxies, typically in a distribution that is round or slightly elongated. Mostly made of dark matter, but can also contain some stars and/or gas.

Hertzsprung-Russell diagram. A plot of star luminosities and spectral types (or equivalent colors or temperatures). Normal stars lie on a "main sequence" running from massive bright blue stars to low-mass dim red stars.

Higgs boson. A fundamental particle thought to be associated with a field that permeates space and explains why particles have mass. The particle was predicted by Peter Higgs and other physicists in 1964, and discovered at the Large Hadron Collider in 2012–13.

lepton. A type of fundamental particle that does not experience the strong nuclear force. The **Standard Model** of particle physics contains six leptons: electrons, **muons**, and **tau** particles, and their associated **neutrinos** (see figure 4.1).

magnification map. A plot showing how much images at different positions are magnified by a cosmic telescope (see figures 6.1 and 6.2).

Massive Astrophysical Compact Halo Objects (MACHOs). Hypothetical dark matter objects made of normal matter, such as **brown dwarfs, white dwarfs, neutron stars**, or **black holes**. The objects are dim or dark but otherwise familiar.

microlensing. A form of gravitational lensing in which a star moves in front of a light source and makes the source appear brighter for a few weeks or months.

millilensing. A form of gravitational lensing in which one or more clumps of dark matter change the observable properties of lensed images.

Modified Newtonian Dynamics (MOND). A hypothesis that attempts to explain cosmic motions by changing Newton's second law of motion from $F = ma$ to $F = ma^2/a_0$ when the acceleration is smaller than a_0.

motion/mass principle. When we see motion induced by

gravity, we can use Newton's law of motion to deduce the force needed to generate the motion, and then use Newton's laws of gravity to determine the amount of mass needed to generate the force.

muon. A fundamental particle similar to the electron but about 200 times more massive.

nebula. Historically, a nebula was any small fuzzy object in the sky. Objects originally labeled *spiral nebulae* are now called **spiral galaxies**. Objects called *planetary nebulae* are now known to be expanding gas clouds produced when a star like the Sun ejects its outer layers near the end of its life. (The name arose because these objects seemed to resemble planets in early observations, but they are not actually related to planets.)

neutrino. A fundamental particle that has very low mass and travels very close to the speed of light. The **Standard Model** of particle physics contains three types of neutrinos (see figure 4.1).

neutron star. A small, extremely dense star that can have about 1.4 times the mass of the Sun but is only about 10 km across. It is the core that remains after a massive star dies and gravity squeezes all of the protons and electrons together to form neutrons.

nuclear fusion. A nuclear reaction in which two atomic nuclei combine to form one larger nucleus. In the Sun, hydrogen undergoes fusion to become helium. Fusion between nuclei less massive than iron releases energy via the famous equation $E = mc^2$.

Ockham's razor. A philosophical principle favoring explanations that are no more complicated than necessary.

photon. The smallest unit of light. According to quantum physics, light acts like both a wave and a particle. A photon is a "particle" of light.

quark. A type of fundamental particle that experiences the strong nuclear force. Quarks combine in pairs or triplets to form various particles (for example, a proton has two up quarks and one down). The **Standard Model of particle physics** contains six types of quarks (see figure 4.1).

quasar. Short for quasi-stellar radio source. An object that looks like a star in the sky but emits as much light as an entire galaxy. The energy is released when matter falls into a **black hole** that is millions or billions of times more massive than the Sun.

rotation curve. A plot of the speed at which stars move in a spiral galaxy, as a function of distance from the center of the galaxy (see figure 3.1).

spectrum. Light spread out like a rainbow so we can see all of the different wavelengths. The full electromagnetic spectrum contains all wavelengths of light. Any observed spectrum shows a finite range of wavelengths (e.g., visible light).

spiral galaxy. A galaxy that has a thin disk with a spiral pattern running through it. The stars in the disk mostly orbit the galaxy in the same direction.

Standard Model of particle physics. A theory that explains the fundamental structure of matter and interactions in terms of twelve matter particles (see figure 4.1), the Higgs boson, and four particles that mediate the electromagnetic force and the weak and strong nuclear forces.

strong lensing. A form of gravitational lensing that creates multiple images and/or giant arcs from individual background sources (cf. **weak lensing**).

subhalo. A clump of dark matter that is bound by its own self-gravity even as it orbits a galaxy. A subhalo that contains both stars and dark matter is a **dwarf galaxy**, while a subhalo that contains only dark matter is a **dark dwarf**.

tau particle. A fundamental particle similar to the electron but about 3,500 times more massive.

tidal force. When two objects interact via gravity, the force on the "front" of each object is a little stronger than the force on the "back." The effect from the Moon creates bulges in Earth's oceans that correspond to the high tides.

time delay. The difference between the amount of time it takes light to reach us for different images in a strong lens system.

weak lensing. A form of gravitational lensing that creates small distortions in the shapes of many background sources (cf. **strong lensing**).

Weakly Interacting Massive Particles (WIMPs). Hypothetical dark matter made of fundamental particles that are different from normal matter. The particles have mass, so they feel gravity, and they interact through the weak nuclear force, but they do not interact with light or participate in nuclear fusion.

white dwarf. A small, dense star that can be as massive as the Sun but only as big as Earth. It is the core that remains after a star like the Sun dies. White dwarfs glow because they are hot, but they no longer produce energy through nuclear fusion.

Notes

Chapter 1   What's in the Dark?

1. The phrase "laws of physics" typically refers to our description of the universe, not the universe itself.

Chapter 3   How Do You Weigh a Galaxy?

1. Van Maanen's claims were later dismissed; astronomers concluded that the measurements were very difficult and that van Maanen thought he saw something that just wasn't there.
2. The situation is different for diffuse nebulae and planetary nebulae, which are in fact "small" gas clouds within the Milky Way.
3. The luminous matter forms a disk because the whole system is spinning. As the protogalactic gas contracts, it spins faster, like a figure skater pulling in her arms. Formally, we say that angular momentum must be conserved, and a disk has the best balance of gravity and angular momentum. This is the same reason that planets tend to lie in a plane around their parent star.

Chapter 4   Is Dark Matter MACHO or WIMPy?

1. As this happens, the Sun will eject its outer layers to form a planetary nebula that will be visible for tens of thousands of years before dissipating into the interstellar medium.
2. There are additional particles associated with the forces (such as the photon for electromagnetism), but they do not enter this discussion.
3. The very young universe was hot and dense enough to be a nuclear reactor.
4. Don't let anyone tell you that physicists are a humorless bunch.

5. One cannot help but notice that no team from the opposing camp called itself the WIMP Project.

CHAPTER 6
"A LONG TIME AGO IN A GALAXY FAR, FAR AWAY"

1. Most of the observing time on the Hubble Space Telescope is allocated through a competitive peer review process, but a small percentage of time is left to the discretion of the director to pursue projects that are worthy but could not necessarily be assigned in the usual way.

Notes on Sources

General

Many theoretical and observational aspects of gravitational lensing are discussed at a technical level in the following reviews:

J. Wambsganss, "Gravitational Lensing in Astronomy," *Living Reviews in Relativity* 1, no. 12 (1998). URL (cited on April 19, 2014): http://www.livingreviews.org/lrr-1998-12.

P. Schneider, C. Kochanek, and J. Wambsganss, *Gravitational Lensing: Strong, Weak and Micro*, Saas-Fee Advanced Course 33, ed. G. Meylan, P. Jetzer, and P. North (New York: Springer, 2006).

Topics in astrophysics that are not explicitly addressed below can be found in textbooks such as:

(Introductory) J. Bennett et al., *The Cosmic Perspective: Fundamentals* (Boston: Addison-Wesley, 2010).

(Introductory) N. F. Comins and W. J. Kaufmann, *Discovering the Universe*, 9th ed. (New York: W. H. Freeman and Company, 2012).

(Advanced) B. W. Carroll and D. A. Ostlie, *An Introduction to Modern Astrophysics*, 2nd ed. (San Francisco: Addison-Wesley, 2006).

(Advanced) C. R. Keeton, *Principles of Astrophysics: Using Gravity and Stellar Physics to Explore the Cosmos* (New York: Springer, 2014).

Preface

The contrast between hubris and humility in cosmology was invoked by Sean Carroll in "Astrophysics: Dark Matter Is Real," *Nature Physics* 2 (2006): 653–54.

"The 'dark' matter surrounds us and penetrates us. . . . It also binds our galaxy together." This is a play on a famous line by Obi-Wan Kenobi in *Star Wars Episode IV.*

CHAPTER 1   WHAT'S IN THE DARK?

For more about spectral classification of stars and the discovery of Neptune, see *Discovering the Universe* by Neil Comins and William Kaufmann (New York: W. H. Freeman and Company, 2012).

Isaac Newton by James Gleick (New York: Pantheon Books, 2003) is a valuable source of information about the man who did much more than teach us about motion and gravity.

The notes for chapter 3 give more information about motions in spiral galaxies.

Fritz Zwicky's analysis of clusters of galaxies is presented in the paper "On the Masses of Nebulae and of Clusters of Nebulae," *The Astrophysical Journal* 86 (1937): 217–46.

Richard Baum and William Sheehan tell the story of Vulcan in their book *In Search of Planet Vulcan: The Ghost in Newton's Clockwork Universe* (New York: Basic Books, 2003).

Mordehai Milgrom's ideas about Modified Newtonian Dynamics were published in three papers:

"A Modification of the Newtonian Dynamics as a Possible Alternative to the Hidden Mass Hypothesis," *The Astrophysical Journal* 270 (1983): 365–70.

"A Modification of the Newtonian Dynamics: Implications for Galaxies," *The Astrophysical Journal* 270 (1983): 371–83.

"A Modification of the Newtonian Dynamics: Implications for Galaxy Systems," *The Astrophysical Journal* 270 (1983): 384–89.

Successes and challenges for MOND are discussed by Benoit Famaey and Stacy McGaugh in their article "Modified Newtonian Dynamics (MOND): Observational Phenomenology and Relativistic Extensions," *Living Reviews in Relativity* 15, no. 10

(2012). URL (cited on April 19, 2014): http://www.livingreviews.org/lrr-2012-10.

Paul Vincent Spade and Claude Panaccio discuss William of Ockham in an entry in *The Stanford Encyclopedia of Philosophy* (Fall 2011 edition), ed. Edward N. Zalta, http://plato.stanford.edu/archives/fall2011/entries/ockham.

Sean Carroll discusses the Standard Model of particle physics and the Higgs boson in *The Particle at the End of the Universe: How the Hunt for the Higgs Boson Leads Us to the Edge of a New World* (New York: Plume, 2013).

Two recent papers present evidence for gamma rays produced by dark matter collisions in the center of the Milky Way (although the interpretation is controversial):

D. Hooper and L. Goodenough, "Dark Matter Annihilation in the Galactic Center as Seen by the Fermi Gamma Ray Space Telescope," *Physics Letters B* 697 (2011): 412–28.

K. N. Abazajian and M. Kaplinghat, "Detection of a Gamma-Ray Source in the Galactic Center Consistent with Extended Emission from Dark Matter Annihilation and Concentrated Astrophysical Emission," *Physical Review D* 86 (2012): id. 083511.

CHAPTER 2   WHEN MASS IS LIKE GLASS

The eclipse expeditions in 1919 are discussed by F. W. Dyson, A. S. Eddington, and C. Davidson in "A Determination of the Deflection of Light by the Sun's Gravitational Field, from Observations Made at the Total Eclipse of May 29, 1919," *Philosophical Transactions of the Royal Society of London*, Series A, vol. 220 (1920): 291–333.

Einstein's analysis of gravitational lensing was published in the paper "Lens-like Action of a Star by the Deviation of Light in the Gravitational Field," *Science* 84 (1936): 506. His cover letter is quoted by J. Renn and T. Sauer in *Revisiting the Foundations of Relativistic Physics*, ed. A. Ashtekar, R. S. Cohen, D. Howard, J. Renn, S. Sarkar, and A. Shimony (Dordrecht: Kluwer, 2003), 69–92.

Fritz Zwicky's suggestion about lensing by "extragalactic nebulae" was published in the paper "Nebulae as Gravitational Lenses," *Physical Review* 51 (1937): 290.

Sjur Refsdal's ideas about gravitational lenses were published in a series of papers:

"The Gravitational Lens Effect," *Monthly Notices of the Royal Astronomical Society* 128 (1964): 295–306.

"On the Possibility of Determining Hubble's Parameter and the Masses of Galaxies from the Gravitational Lens Effect," *Monthly Notices of the Royal Astronomical Society* 128 (1964): 307–10.

"On the Possibility of Testing Cosmological Theories from the Gravitational Lens Effect," *Monthly Notices of the Royal Astronomical Society* 132 (1966): 101–11.

Here are key papers reporting discoveries of the first two gravitational lens systems:

D. Walsh, R. F. Carswell, and R. J. Weymann, "0957+561 A, B: Twin Quasistellar Objects or Gravitational Lens?" *Nature* 279 (1979): 381–84.

P. Young et al., "The Double Quasar Q0957+561 A, B: A Gravitational Lens Image Formed by a Galaxy at $Z = 0.39$," *The Astrophysical Journal* 241 (1980): 507–20.

R. J. Weymann et al., "The Triple QSO PG1115+08: Another Probable Gravitational Lens," *Nature* 285 (1980): 641–43.

The image shown in figure 2.11 comes from a paper by C. R. Keeton et al., "Differential Microlensing of the Continuum and Broad Emission Lines in SDSS J0924+0219, the Most Anomalous Lensed Quasar," *The Astrophysical Journal* 639 (2006): 1–6.

Paul Schechter describes gravitational lensing in terms of the three D's (deflection, distortion, and delay) in "H_0 from Gravitational Lenses: Recent Results," *New Cosmological Data and the Values of the Fundamental Parameters (IAU Symposium #201)*, ed. A. Lasenby and A. Wilkinson (San Francisco: Astronomical Society of the Pacific, 2005), 209–18.

CHAPTER 3 HOW DO YOU WEIGH A GALAXY?

The story of van Maanen's claims about the rotation of spiral nebulae is discussed by Norriss Hetherington in "Adrian van Maanen and Internal Motions in Spiral Nebulae: A Historical Review," *Quarterly Journal of the Royal Astronomical Society* 13 (1972): 25–39.

The arguments from the "Great Debate" were published by Harlow Shapley and Heber Curtis in "The Scale of the Universe," *Bulletin of the National Research Council* 2, pt. 3, no. 11 (1921): 171–217.

Edwin Hubble's work on "spiral nebulae" is discussed in his book *The Realm of the Nebulae* (1936; rpt. New Haven: Yale University Press, 2013).

For investigations of spiral galaxy rotation curves, see:

V. C. Rubin, W. K. Ford Jr., and N. Thonnard, "Extended Rotation Curves of High-Luminosity Spiral Galaxies. IV. Systematic Dynamical Properties, Sa-Sc," *The Astrophysical Journal Letters* 225 (1978): L107–L111.

M. S. Roberts, "The Rotation Curves of Galaxies," *Comments on Astrophysics* 6 (1976): 105–10.

The sources for chapter 1 give references about Modified Newtonian Dynamics.

Several papers use dwarf galaxies to study galaxy mass distributions:

D. Zaritsky, R. Smith, C. Frenk, and S. D. M. White, "More Satellites of Spiral Galaxies," *The Astrophysical Journal* 478 (1997): 39–48.

T. A. McKay et al., "Dynamical Confirmation of Sloan Digital Sky Survey Weak-Lensing Scaling Laws," *The Astrophysical Journal Letters* 571 (2002): L85–L88.

F. Prada et al., "Observing the Dark Matter Density Profile of Isolated Galaxies," *The Astrophysical Journal* 598 (2003): 260–71.

Ortwin Gerhard reviews methods for studying the mass distributions of elliptical galaxies in the article "Dark Matter in Massive Galaxies," *The Intriguing Life of Massive Galaxies (IAU Symposium #295)*, ed. D. Thomas, A. Pasquali, and I. Ferreras (Cambridge: Cambridge University Press, 2013), 211–20.

Tommaso Treu discusses various ways in which gravitational lensing can be combined with other information to study galaxy mass distributions in his review article "Strong Lensing by Galaxies," *Annual Reviews of Astronomy and Astrophysics* 48 (2010): 87–125.

Ann Zabludoff and I examine how a galaxy's neighbors affect gravitational lensing in our paper "The Importance of Lens Galaxy Environments," *The Astrophysical Journal* 612 (2004): 660–78.

Joshua Barnes and Lars Hernquist describe what happens when galaxies collide in "Dynamics of Interacting Galaxies," *Annual Reviews of Astronomy and Astrophysics* 30 (1992): 705–42.

CHAPTER 4   IS DARK MATTER MACHO OR WIMPY?

A famous test of Einstein's theory of relativity involving muons was reported by D. H. Frisch and J. H. Smith in "Measurement of the Relativistic Time Dilation Using μ-Mesons," *American Journal of Physics* 31 (1963): 342–55.

The prediction of anti-electrons was presented by Paul Dirac in "Quantised Singularities in the Electromagnetic Field," *Proceedings of the Royal Society of London*, Series A, vol. 133 (1931): 60–72. The discovery of anti-electrons was reported by Carl Anderson in "The Positive Electron," *Physical Review* 43 (1933): 491–94.

The discovery of the top quark was reported by F. Abe et al. in "Observation of Top Quark Production in $\bar{p}p$ Collisions with the Collider Detector at Fermilab," *Physical Review Letters* 74 (1995): 2626–31.

The discovery of the tau neutrino was reported by K. Kodama et al. in "Observation of Tau Neutrino Interactions," *Physics Letters B* 504 (2001): 218–24.

Timothy J. Sumner discusses particle physics aspects of dark matter in his article "Experimental Searches for Dark Matter,"

Living Reviews in Relativity 5, no. 4 (2002). URL (cited on April 19, 2004): http://www.livingreviews.org/lrr-2002-4.

The acronym MACHO was coined by Kim Griest in "Galactic Microlensing as a Method of Detecting Massive Compact Halo Objects," *The Astrophysical Journal* 366 (1991): 412–21. Bohdan Paczyński proposed to use gravitational microlensing to look for MACHOs in the paper "Gravitational Microlensing by the Galactic Halo," *The Astrophysical Journal* 304 (1986): 1–5. The following papers report results from microlensing surveys:

C. Alcock et al., "The MACHO Project: Microlensing Optical Depth toward the Galactic Bulge from Difference Image Analysis," *The Astrophysical Journal* 541 (2000): 734–66.

C. Alcock et al., "The MACHO Project: Microlensing Results from 5.7 Years of Large Magellanic Cloud Observations," *The Astrophysical Journal* 542 (2000): 281–307.

A. Udalski, M. Kubiak, and M. Szymański, "Optical Gravitational Lensing Experiment. OGLE-2—The Second Phase of the OGLE Project," *Acta Astronomica* 47 (1997): 319–44.

P. Tisserand et al., "Limits on the Macho Content of the Galactic Halo from the EROS-2 Survey of the Magellanic Clouds," *Astronomy and Astrophysics* 469 (2007): 387–404.

Primordial abundances of light elements are discussed by Achim Weiss in "Elements of the Past: Big Bang Nucleosynthesis and Observation," *Einstein Online* 2 (2006): 1019.

The statement about WIMPs traveling through a solar system's worth of lead is based on the putative WIMP properties reported in the paper by Agnese et al. (2013) listed below.

For a general introduction to neutrinos, see Frank Close, *Neutrino* (Oxford: Oxford University Press, 2010). For an advanced review of neutrino oscillations, see M. C. Gonzalez-Garcia and Y. Nir, "Neutrino Masses and Mixing: Evidence and Implications," *Reviews of Modern Physics* 75 (2003): 345–402.

Raymond Davis's neutrino detector was described in an article by John Bahcall, "Neutrinos from the Sun," *Scientific American* 221 (1969): 29–37. Results from recent studies of solar neutrinos are given in several papers, including:

J. Hosaka et al., "Solar Neutrino Measurements in Super-Kamio-kande-I," *Physical Review D* 73 (2006): id. 112001.

Q. R. Ahmad et al., "Direct Evidence for Neutrino Flavor Transformation from Neutral-Current Interactions in the Sudbury Neutrino Observatory," *Physical Review Letters* 89 (2002): id. 011301.

The idea that Earth's motion around the Sun affects WIMP searches was initially discussed in two papers:

A. K. Drukier, K. Freese, and D. N. Spergel, "Detecting Cold Dark-Matter Candidates," *Physical Review D* 33 (1986): 3495–508.

K. Freese, J. Frieman, and A. Gould, "Signal Modulation in Cold-Dark-Matter Detection," *Physical Review D* 37 (1988): 3388–405.

Results from WIMP searches are reported in the following papers:

R. Bernabei et al., "New Results from DAMA/LIBRA," *The European Physical Journal C* 67 (2010): 39–49.

R. Agnese et al., "Silicon Detector Dark Matter Results from the Final Exposure of CDMS II," *Physical Review Letters* 111 (2013): id. 251301.

C. E. Aalseth et al., "CoGeNT: A Search for Low-Mass Dark Matter Using P-Type Point Contact Germanium Detectors," *Physical Review D* 88 (2013): id. 012002.

G. Angloher et al., "Results from 730 kg Days of the CRESST-II Dark Matter Search," *The European Physical Journal C* 72 (2012): id. 1971.

E. Aprile et al., "Dark Matter Results from 225 Live Days of XENON100 Data," *Physical Review Letters* 109 (2012): id. 181301.

D. S. Akerib et al., "First Results from the LUX Dark Matter Experiment at the Sanford Underground Research Facility," *Physical Review Letters*, 112 (2014): id. 091303.

CHAPTER 5   FINDING WHAT'S MISSING

Here are examples of papers presenting models for early samples of gravitational lenses:

C. S. Kochanek, "The Implications of Lenses for Galaxy Structure," *The Astrophysical Journal* 373 (1991): 354–68.

C. R. Keeton, C. S. Kochanek, and U. Seljak, "Shear and Ellipticity in Gravitational Lenses," *The Astrophysical Journal* 482 (1997): 604–20.

The image of the four-image gravitational lens B1555+375 in figure 5.1 comes from D. R. Marlow et al., "CLASS B1555+375:

A New Four-Image Gravitational Lens System," *The Astronomical Journal* 118 (1999): 654–58.

R. W. Schmidt and J. Wambsganss discuss cosmological microlensing in their review article "Quasar Microlensing," *General Relativity and Gravitation* 42 (2010): 2127–50.

The paper by Shude Mao and Peter Schneider is "Evidence for Substructure in Lens Galaxies?" *Monthly Notices of the Royal Astronomical Society* 295 (1998): 587–94.

The simulated galaxy shown in figure 5.2 comes from J. Diemand et al., "Clumps and Streams in the Local Dark Matter Distribution," *Nature* 454 (2008): 735–38.

The "missing satellites problem" was originally identified in two papers:

B. Moore et al., "Dark Matter Substructure within Galactic Halos," *The Astrophysical Journal Letters* 524 (1999): L19–L22.
A. Klypin et al., "Where Are the Missing Galactic Satellites?" *The Astrophysical Journal* 522 (1999): 82–92.

Andrey Kravtsov discusses the problem and possible solutions in "Dark Matter Substructure and Dwarf Galactic Satellites," *Advances in Astronomy* (2010): id. 281913.

Here are the original papers about lensing and dark matter substructure:

R. B. Metcalf and P. Madau, "Compound Gravitational Lensing as a Probe of Dark Matter Substructure within Galaxy Halos," *The Astrophysical Journal* 563 (2001): 9–20.
M. Chiba, "Probing Dark Matter Substructure in Lens Galaxies," *The Astrophysical Journal* 565 (2002): 17–23.
N. Dalal and C. S. Kochanek, "Direct Detection of Cold Dark Matter Substructure," *The Astrophysical Journal* 572 (2002): 25–33.

Here are papers about fitting lenses with more general galaxy shapes:

N. W. Evans and H. J. Witt, "Fitting Gravitational Lenses: Truth or Delusion," *Monthly Notices of the Royal Astronomical Society* 345 (2003): 1351–64.
A. B. Congdon and C. R. Keeton, "Multipole Models of Four-Image

Gravitational Lenses with Anomalous Flux Ratios," *Monthly Notices of the Royal Astronomical Society* 364 (2005): 1459–66.

J. Yoo et al., "The Lens Galaxy in PG 1115+080 Is an Ellipse," *The Astrophysical Journal* 626 (2005): 51–57.

J. Yoo et al., "Halo Structures of Gravitational Lens Galaxies," *The Astrophysical Journal* 642 (2006): 22–29.

Here are papers about the connection between mathematics and lensing:

C. R. Keeton, B. S. Gaudi, and A. O. Petters, "Identifying Lenses with Small-Scale Structure. I. Cusp Lenses," *The Astrophysical Journal* 598 (2003): 138–61.

C. R. Keeton, B. S. Gaudi, and A. O. Petters, "Identifying Lenses with Small-Scale Structure. II. Fold Lenses," *The Astrophysical Journal* 635 (2005): 35–59.

Here are papers about mass clumps in individual quasar lenses:

J. P. McKean et al., "High-Resolution Imaging of the Anomalous Flux Ratio Gravitational Lens System CLASS B2045+265: Dark or Luminous Satellites?" *Monthly Notices of the Royal Astronomical Society* 378 (2007): 109–18.

A. More et al., "The Role of Luminous Substructure in the Gravitational Lens System MG 2016+112," *Monthly Notices of the Royal Astronomical Society* 394 (2009): 174–90.

R. M. Kratzer et al., "Analyzing the Flux Anomalies of the Large-Separation Lensed Quasar SDSS J1029+2623," *The Astrophysical Journal Letters* 728 (2011): id. L18.

R. Fadely and C. R. Keeton, "Substructure in the Lens HE 0435–1223," *Monthly Notices of the Royal Astronomical Society* 419 (2012): 936–51.

M. Oguri et al., "The Hidden Fortress: Structure and Substructure of the Complex Strong Lensing Cluster SDSS J1029+2623," *Monthly Notices of the Royal Astronomical Society* 429 (2013): 482–93.

C. L. MacLeod et al., "Detection of Substructure in the Gravitationally Lensed Quasar MG0414+0534 Using Mid-infrared and Radio VLBI Observations," *The Astrophysical Journal* 773 (2013): id. 35.

Here are similar papers about ring lenses:

S. Vegetti et al., "Detection of a Dark Substructure through Gravitational Imaging," *Monthly Notices of the Royal Astronomical Society* 408 (2010): 1969–81.

S. Vegetti et al., "Gravitational Detection of a Low-Mass Dark Satellite Galaxy at Cosmological Distance," *Nature* 481 (2012): 341–43.

Here are theoretical papers about substructure lensing:

J. Chen et al., "Astrometric Perturbations in Substructure Lensing," *The Astrophysical Journal* 659 (2007): 52–68.

C. R. Keeton and L. A. Moustakas, "A New Channel for Detecting Dark Matter Substructure in Galaxies: Gravitational Lens Time Delays," *The Astrophysical Journal* 699 (2009): 1720–31.

Here are papers that discuss prospects for gravitational lensing with new astronomical surveys:

L. V. E. Koopmans et al., "Strong Gravitational Lensing as a Probe of Gravity, Dark-Matter and Super-Massive Black Holes," science white paper submitted to Astro2010: Astronomy and Astrophysics Decadal Survey (2009), electronic preprint permanently archived at http://arxiv.org/abs/0902.3186.

P. J. Marshall et al., "Dark Matter Structures in the Universe: Prospects for Optical Astronomy in the Next Decade," science white paper submitted to Astro2010: Astronomy and Astrophysics Decadal Survey (2009), electronic preprint permanently archived at http://arxiv.org/abs/0902.2963.

L. A. Moustakas et al., "Strong Gravitational Lensing Probes of the Particle Nature of Dark Matter," science white paper submitted to Astro2010: Astronomy and Astrophysics Decadal Survey (2009), electronic preprint permanently archived at http://arxiv.org/abs/0902.3219.

M. Oguri and P. J. Marshall, "Gravitationally Lensed Quasars and Supernovae in Future Wide-Field Optical Imaging Surveys," *Monthly Notices of the Royal Astronomical Society* 405 (2010): 2579–93.

Y. Hezaveh et al., "Dark Matter Substructure Detection Using Spatially Resolved Spectroscopy of Lensed Dusty Galaxies," *The Astrophysical Journal* 767 (2013): id. 9.

For more information about the Large Synoptic Survey Telescope, see http://www.lsst.org/lsst.

CHAPTER 6
"A LONG TIME AGO IN A GALAXY FAR, FAR AWAY"

The title of this chapter is taken from *Star Wars*, of course.

For more information about the Cosmic Microwave Background, see:

(General) George Smoot and Keay Davidson, *Wrinkles in Time: Witness to the Birth of the Universe* (New York: Harper Perennial, 2007).

(Technical) A. Jones and A. N. Lasenby, "The Cosmic Microwave

Background," *Living Reviews in Relativity* 1, no. 11 (1998). URL (cited on April 19, 2014): http://www.livingreviews.org/lrr-1998-11.

The various Hubble deep field observations are reported in the following papers:

R. E. Williams et al., "The Hubble Deep Field: Observations, Data Reduction, and Galaxy Photometry," *The Astronomical Journal* 112 (1996): 1335–89.

S. V. W. Beckwith et al., "The Hubble Ultra Deep Field," *The Astronomical Journal* 132 (2006): 1729–55.

G. D. Illingworth et al., "The HST eXtreme Deep Field (XDF): Combining All ACS and WFC3/IR Data on the HUDF Region into the Deepest Field Ever," *The Astrophysical Journal Supplement Series* 209 (2013): id. 6.

For more information about the new Hubble Space Telescope campaign to observe fields that include cosmic telescopes, see "Hubble Boldly Goes: The Frontier Fields Program," *Newsletter of the Space Telescope Science Institute* 30, no. 1 (2013), 1–6; also see http://www.stsci.edu/hst/campaigns/frontier-fields.

Steven Allen et al. discuss galaxy clusters and cosmology in their review article "Cosmological Parameters from Observations of Galaxy Clusters," *Annual Review of Astronomy and Astrophysics* 49 (2011): 409–70.

Jean-Paul Kneib and Priyamvada Natarajan discuss many aspects of cluster lensing, including lens modeling and weak lensing, in their review article "Cluster Lenses," *The Astronomy and Astrophysics Review* 19 (2011): id. 47.

Here are papers from our research team about finding outstanding cosmic telescopes:

K. C. Wong et al., "Optimal Mass Configurations for Lensing High-Redshift Galaxies," *The Astrophysical Journal* 752 (2012): id. 104.

K. C. Wong et al., "A New Approach to Identifying the Most Powerful Gravitational Lensing Telescopes," *The Astrophysical Journal* 769 (2013): id. 52.

S. M. Ammons et al., "Mapping Compound Cosmic Telescopes Containing Multiple, Projected Cluster-Scale Halos," *The Astrophysical Journal* 781 (2014): id. 2.

K. D. French et al., "Characterizing the Best Cosmic Telescopes with the Millennium Simulations," *The Astrophysical Journal*, 785 (2014): id. 59.

Key studies of the Bullet Cluster system are presented in the following papers:

M. Markevitch et al., "A Textbook Example of a Bow Shock in the Merging Galaxy Cluster 1E 0657-56," *The Astrophysical Journal Letters* 567 (2002): L27–L31.

D. Clowe et al., "A Direct Empirical Proof of the Existence of Dark Matter," *The Astrophysical Journal Letters* 648 (2006): L109–L113.

Here are some papers about line-of-sight effects in cluster lensing:

E. Jullo et al., "Cosmological Constraints from Strong Gravitational Lensing in Clusters of Galaxies," *Science* 329 (2010): 924–27.

A. D'Aloisio and P. Natarajan, "Cosmography with Cluster Strong Lenses: The Influence of Substructure and Line-of-Sight Haloes," *Monthly Notices of the Royal Astronomical Society* 411 (2011): 1628–40.

O. Host, "Galaxy Cluster Strong Lensing: Image Deflections from Density Fluctuations along the Line of Sight," *Monthly Notices of the Royal Astronomical Society* 420 (2012): L18–L22.

A. D'Aloisio, P. Natarajan, and P. R. Shapiro, "The Effect of Large-Scale Structure on the Magnification of High-Redshift Sources by Cluster-Lenses," *Monthly Notices of the Royal Astronomical Society*, submitted, electronic preprint permanently archived at http://arxiv.org/abs/1311.1614.

About the Author

CHARLES KEETON is an associate professor of physics and astronomy at Rutgers University. He has worked with the Hubble Space Telescope and observatories in Arizona and Chile, and published more than ninety articles in astronomy journals. In 2010, Dr. Keeton received the Presidential Early Career Award for Scientists and Engineers from President Barack Obama.